THE MATRIX REFORMED
Science Fiction, Technology,
and Christian Philosophy

THE MATRIX REFORMED
Science Fiction, Technology, and Christian Philosophy

Bart Cusveller
Maarten Verkerk
Marc de Vries

DORDT COLLEGE PRESS

Cover design by Rob De Haan
Copy layout by Carla Goslinga

Copyright © 2011 by Bart Cusveller

This book is a translation from the Dutch by Renate Leder of *De Matrix Code. Sciencefictionfilms als spiegel van de technologische cultuur*, Buijten & Schipperheijn, Amsterdam, 2006.

Fragmentary portions of this book may be freely used by those who are interested in sharing these authors' insights and observations, so long as the material is not pirated for monetary gain and so long as proper credit is visibly given to the publisher, the authors, and the translator. Others, and those who wish to use larger sections of text, must seek written permission from the publisher.

Printed in the United States of America.

Dordt College Press www.dordt.edu/dordt_press
498 Fourth Avenue NE
Sioux Center, Iowa 51250
United States of America

ISBN 978-0-932914-90-3

Library of Congress Cataloging-in-Publication Data: 2011933199

CONTENTS

Foreword 1

Preface 3

1. The Fantastic Flight of Science Fiction 7
2. Watching Movies from a Christian Perspective 19
3. Reformational Philosophy Reloaded 27
4. *The Matrix*: A Modern Myth? 39
5. The Lion, the Witch, and the Matrix 49
6. Technology as Threat 59
7. Technology as Utopia 71
8. Technology between Reality and Perception 81
9. Illusion and Reality 91
10. Freedom, Determinism, or Destiny? 101
11. Can Programs make Moral Judgments? 113
12. Morality in a High-tech Society 125
13. Belief and Rationality 137

Conclusion 147

Bibliography 149

Foreword

It gives me great pleasure to write a brief foreword to this work by Bart Cusveller, Maarten Verkerk, and Marc de Vries.

Those who are familiar with the scope and practical relevance of Reformational philosophy will not be surprised to see how this philosophy opens up a fresh understanding of the intertwinement of various facets of our contemporary culture, focused on science fiction films while contemplating the interplay of philosophy, culture, and film.

What is normally taken for granted—namely, our contemporary technological worldview—is brought to the surface in its magnified form by science fiction as a genre. The authors ask many burning questions and do not hesitate to explore their answers in various directions, such as by contemplating opposing appreciations of technology (as a threat or utopia, positioned between reality, perception, and illusion) by considering moral issues and by watching movies from a Christian perspective.

One of the underlying questions is which techniques and technologies are compatible with what is good for human beings. This relates to the aim that the authors have set for themselves: "When the reader finishes reading all of this, he will have been provided food for thought on aspects of modern culture (as these are urgent), on Reformational philosophy (as it is helpful), and on science fiction films (as they offer a good starting point for our reflections)."

This aim is successfully achieved in a richly nuanced way by relating the issues under discussion with a diversity of science fiction films. It also investigates how Christians ought to respond to such films, whether through condemnation, appropriation, consumption, or distinction. Various distinctions found in Reformational philosophy are explored in relation to the theme and literature relevant to science fiction is brought into the discussions.

On the whole, this work is thought-provoking; it calls for us to reconsider much of what we are implicitly (and oftentimes uncritically) accepting. Whoever reads it will feel enriched by multiple constructive and elucidating distinctions and insights, well aware of the fact that it opened up a deepened perspective on our culture and the role of science fiction films within it.

D.F.M.Strauss
Bloemfontein, South Africa – March 2011

PREFACE

> Morpheus: "Take the red pill and I will show you
> how deep the rabbit hole goes."
> (*The Matrix*)

This book revolves around the intersection of three themes: science fiction, technology, and philosophy. Our main focus is on the analysis of the technological culture as it is represented in science fiction films. For this analysis, we will draw on Reformational philosophy.

The immediate cause of this book is that two of the authors have had a positive experience with the use of science fiction films for teaching philosophical themes concerning technology and technological culture to students of two technological universities. Technology in itself and, in a broader sense, a technological worldview have become so common that its self-evidence is only revealed and cracked in the extreme magnification of it in a genre like science fiction. This is all the more true when represented by visual media like TV, film, and computer games. And when something loses its self-evidence and is replaced by amazement and perhaps bewilderment, philosophers can come into action. However, as there was no constructive collection of texts available for philosophical reflection on the intersection between science fiction, technology, and philosophy, the authors have written this book on the subject.

The theme of this book originates from the fact that the authors find themselves in the middle of a field of thoughts and questions on philosophy, culture, and film. It has been said, for example, that we are living in an image culture. What seems to count in our culture is what looks good and can be visually presented. The human body is accordingly adapted (see commercials), the value of devices used daily are increasingly determined by visual applications (see cell phones with camera applications), and organizations are judged by recognizable styling (see website presentation). In the meantime, it is no longer noticed that all this visual violence is made possible by technological means and by a way of life that is directed at technological possibilities. The culture surrounding us is, in many different ways, drenched in an ever-increasing inundation of technology, and this is poignantly expressed in science fiction.

Many of us experience a certain kind of discontent at this inundation of technology. Before you know how the current computer works, it has already become outdated. At the birth of a child, a father stands not next to his wife but behind a digital camera. A company is not successful if it does not have a good commercial on TV. Churches are using digital projectors during sermons and worship. Adults want to use webcams while chatting, and children may not feel accepted until they have the latest computer game. Or is it just unfamiliarity, which might mean we will eventually grow used to the product in the end? Perhaps the source of discontent is precisely that we do not know how to *differentiate* between feelings of alienation from reality by the high-technological degree of our culture and feelings of unfamiliarity that will subside in the end. Due to all the fundamental changes in our culture and in our orientations on that culture, we hardly know how to decide which is which.

We could also formulate this as a question: where can we find truly human characteristics that tell us what fits with our life and what alienates us from it? What techniques and technologies are compatible with what is good for a human being? Such questions are urgently addressed in relation to the medical possibilities (e.g., cloning, embryo culture, genetic manipulation, and many more), but also in relation to agriculture, the environment, recreation, industry, and other sectors of our society. And it is often in science fiction that such possibilities are strongly addressed. Suppose we can clone a person; what does that say about the identity of this newly created person? Suppose we can alter a person's memory; what would that mean for that person's happiness? Suppose our culture is destroyed by nuclear disaster; what kind of people would we turn out to be? Suppose we could colonize space; would we have something good to offer? The subject matter is nearly inexhaustible. That is why such films offer a very useful starting point for reflection on the discontent of the effects of technological culture on the possibilities to live a human life. We will tackle a number of these issues in this book.

Our goal is to show that many of these questions are fundamentally linked to the way people look at life, what their deepest orientation on reality is, and what they believe about the origin, purpose, and meaning of their existence. Their morality, their philosophy, makes a difference for the valuation and directing of technology and technological culture. That may sound as if the authors of this book can take a distant look at all of this. But the truth is, of course, that nobody can examine interpretations of fundamental givens of human existence from a morally neutral or philosophically impartial viewpoint. In fact, we as authors find it important explicitly *not* to do that. We do not only stand right in the middle of contemporary culture, but we are also part of a philosophical tradition, the Reformational tradition of Christianity, which we consider to be essential for thinking through and valuating the starting points, phenomena, and effects of technologization in modern culture. By taking this approach, we hope to be helpful to other Christians in their attitude towards certain social concerns in our modern culture. A not unimportant side effect is that, hopefully, people will look more seriously from a Christian angle at expressions of popular visual media, in *casu* film. Under the surface of films, TV, and computer games there is often plenty of tossing and turning that is philosophically relevant and deserves attention.

In that respect, we, the authors, are among others inspired by a body of thought also referred to as Reformational philosophy. Exercising philosophy has an analytical, critical, and orientating function. Reformational philosophy is a form of Christian thinking that has been developed and expanded in the footprints of two Dutch thinkers, Herman Dooyeweerd and Dirk Vollenhoven. What they originally termed as the "philosophy of the law-idea" was taken up by students and interested people and was sometimes applied in a stricter, sometimes in a softer sense for discussing subjects other than they had intended. Medical science is such a subject, but so is science fiction! Who would ever have imagined Herman Dooyeweerd would be mentioned in one sentence with Arnold Schwarzenegger, and Henk Geertsema with Obi-Wan Kenobi? In the discussion of technological cultures as we see it in science fiction films, a number of elements from Reformational philosophy are introduced that shed light on our topic. Think of the notions that perspectives on society are never morally and philosophically impartial, that reality has a fundamentally multifaceted and predefined order, and that human action always entails the responsibility to disclose a multitude of norms. In that sense, this book also lends access to this form of Christian thought.

When the reader finishes reading all of this, he will have been provided food for thought on aspects of modern culture (as these are urgent), on Reformational philosophy (as it is helpful), and on science fiction films (as they offer a good starting point for our reflections). Technological culture involves real questions and problems, and we believe that Reformational philosophy offers a well-balanced insight into—and perspective on—our time. The reader will have an approach route via a number of science fiction films. The gist is that science fiction films are not only beautifully made and fun to watch, but also informative.

The field we have mapped out above will be explored in this book in thirteen short chapters, roughly subdivided into three groups. First of all, we invite the reader on our journey through time and space in a number of chapters where our access road of science fiction and science fiction films are explained. The trilogy of *The Matrix*, in particular, forms a focal point but other films and series are also referred to, such as *Star Trek* and *Star Wars*. In addition, we will look at several other philosophers that draw on the same discourse. The official website of *The Matrix*, for example, included a section where professional philosophers posted their reflections (now published with classical texts by Oxford University Press[1]). Then, in connection to the initial explorations, we will concentrate on a number of philosophical themes that are brought forward by the technological world and her problems, like the ones presented by science fiction. The philosophy of technology and the technological culture also have their specific body of texts that the authors will take into account. The last few chapters take the reader deeper into topics that can be addressed from a number of fundamental philosophical disciplines. Then it is time to assess the results and return to the thesis from this preface.

It will not be a surprise that the three authors are both Christian philosophers and science fiction film fans. In daily life, Bart Cusveller is professor of nursing ethics at the School of Nursing, Christian University for Applied Sciences, Ede, The Netherlands; Maarten Verkerk is affiliate professor of Reformational Philosophy at the University of Eindhoven and Maastricht; and Marc de Vries is affiliate professor Reformational Philosophy at the University of Delft. Their appearance in alphabetical order does not necessarily indicate an order of the effort put into the book.

To conclude, we wish to thank those with whom we spent countless hours watching films, and friends and colleagues who read and commented on (parts of) an early draft of the manuscript.

We also would like to thank Renate Leder for the translation of the book. She did a very good job. Her translated text is sometimes more fluent than the original text! To them and all other readers: *Live long and prosper!*[2]

---

1   Christopher Grau (ed.), *Philosophers Explore* The Matrix (Oxford: Oxford University Press, 2005).

2   "Live long and prosper" is the traditional salutation given by the Vulcan Dr. Spock in the science fiction series *Star Trek*, taken by the makers from the Aaronic blessing.

Chapter 1

THE FANTASTIC FLIGHT OF SCIENCE FICTION

> Roy Batty: "I've seen things you people wouldn't believe. Attack ships on fire on the shoulder of Orion. I watched sea beams glitter in the dark near the Tannhauser Gate."
> (*Blade Runner*)

1. Science Fiction: "Free your mind"

This book revolves around the question "what can a genre of popular films like science fiction tell us about the culture we live in?" This will be discussed in greater detail in the following chapters. But first, what exactly is science fiction? What renders this type of film suitable for discussion here? In this chapter, we will provide some preliminary observations on science fiction as a genre in literature and film.

Science fiction and its cousin, fantasy, have enjoyed enormous popularity. Not so long ago, cinemas showed the final parts of the best-selling trilogies *Star Wars*, *The Lord of the Rings*, and *The Matrix* to rave reviews. In 2009 and 2010, *Avatar* made more money at the Box Office and in movie rentals than any other movie in history: $2.78 billion dollars.[1] Polls conducted in Australia and the United Kingdom showed tens of thousands declaring to be followers of the Jedi doctrine (whether or not as a joke). In terms of statistics, the Jedi belief system is sufficient to be counted as an official religion. At least a dozen science fiction series are currently shown on Dutch television. The United States and the United Kingdom even have special TV channels, magazines, and books on the genre. And science fiction-themed websites and computer games abound ever since the digital revolution. Indeed, science fiction as a cultural phenomenon has taken a fantastic flight.

This brings a tremendous diversity in style: each book and website differs from the next, and science fiction movies digitally range from

---

1 www.imdb.com/boxoffice/alltimegross?region=world-wide (retrieved March 19, 2011).

standard classics to substandard attempts to problematic borderline cases.

We will focus on the historical dimensions of the genre below, but first we need to address the question whether it is possible to point out "family resemblance" within works of science fiction.

2. "It's the only way to fly"

The genre of science fiction, or SF, is often associated with stories set in the future or in space. They are often combined with a bit of supernatural flying and crashing and a couple of strange, unearthly creatures. But having these elements in a movie does not necessarily make it science fiction. There are also space and future movies whose stories may as well have been set in the Wild West, in prehistory, or in Germany during the Second World War. Fans of the "real" science fiction mockingly refer to these as "space opera."

So what defines "real" science fiction? Let us start at the beginning, with the name of the genre. The term *science fiction* literally refers to an invented development or event ("fiction") in science and/or technology ("science") and indicates that this development or event is central to the plot. "Real" science fiction is therefore not just any story, however fantastic or futuristic, but rather revolves around a thought experiment ("what if. . .") about the possibilities opening up through developments in scientific knowledge and technological know-how. This could either be because of our progressing scientific and technological development (*Gattaca*, for example, features genetic manipulation of humans), or because we come into contact with a civilization that is more advanced in terms of science and technology (*Close Encounters of the Third Kind*). It is that development or event, either from within our world or from outside, that creates a situation distinctly different from what is possible today or what was possible in the past. Such stories are science fiction in the narrow sense of the word: an extrapolating thought experiment about science and technology that confronts us with an unfamiliar situation.[2] Space and future will often, but not invariably, play a role. A case in point is *Contact*, a film that takes place in the here and now, when an extraterrestrial intelligence makes contact. Another example is *Sky Captain and the World of Tomorrow*; this movie is set in the past, but centers on a development or event that is impossible in terms of our current knowledge and know-how.

---

2   Cf. Deborah Knight, George McKnight, "Real Genre and Virtual Philosophy," in William Irvin (ed.), *The Matrix and Philosophy* (Chicago: Open Court, 2002) 194; see also "Science fiction" and "Science fiction film," www.wikipedia.com (retrieved 14 March 2011).

Some of these films are intended for entertainment purposes, as a fantastical escape from the daily rut. Science fiction films featuring Arnold Schwarzenegger guarantee a decent amount of heroic and violent action for the cinema audience to enjoy. For other films, the thought experiment mentioned before is a more serious matter. The filmmaker wishes to make something artistic (consider *2001: A Space Odyssey*, or *Solaris*), and to explore an idea or line of thought. The aim of such films is not to entertain, but to express something essential. It is fair to say in most cases that science fiction films try to do both. *Inception* is a good example of a movie that is intended for the audience to think as well as to be entertained.

At any rate, the significance of a fictional scientific or technological development for culture and society in general, or for the protagonists in particular, plays a central part in this kind of science fiction.

As may be expected, this gives rise to a number of questions concerning definition and interpretation: what films belong to this genre, and which do not? This problem often arises when a definition is given first, and the examples are provided afterwards. So we could reverse the order and first give an example that most people would agree belongs to the core of classics in the genre, and only then try to define that core.[3] We will see that elements from the description given above indeed do occur. From there, we can discuss the cases that, due to analogies and similarities, belong to the periphery of borderline cases.

One such classic is certainly Ridley Scott's *Blade Runner* (1984), based on a story by Philip K. Dick ("Do androids dream of electric sheep?"). The movie shows a gloomy future where it is possible to create biologically sound but artificial copies of humans. These creatures, called "androids" in the novel and "replicants" in the film, are stationed in extraterrestrial colonies to make life easier for human beings. They are endowed with artificial memories as a frame for their consciousness. In order to control these androids' desires and emotions, they also have a built-in genetic condition, limiting their lifespan to four years. However, a few of these replicants are angered when they start to see what real humans have: a real past and a real future, with real love and real hope. When they decide to seek justification with their creator, they are hunted down by the main character, "blade runner" Deckard, an agent with "a license to kill." The suspense in the movie almost distracts the viewer from

---

3   Mark Rowlands, *The Philosopher at the End of the Universe: Philosophy explained through science fiction film* (London: Ebury, 2003) ix (previously published as *Sci-Phi: Philosophy from Socrates to Schwarzenegger*) 233.

the questions surrounding the protagonist: what makes his life more human than that of the replicants? Indeed, could he not be, as the original story and the director's cut suggest, a replicant himself? If he is, what is the distinguishing factor, and what is the value of such a distinction? That question can only really be posed in the world of *Blade Runner*, but at the same time, it has wider relevance: what makes humans human, and what is the significance of authenticity of memory and desire? According to us, such a film is defining for the SF genre.

3. Back to the future: from Frankenstein to Neo

In this chapter we want to examine a few versions and subgenres of science fiction, and subsequently consider its core in greater detail as this is significant for our interest in *The Matrix*. But before we do so, it is important to first discuss the historical development of the literary and cinematic genre of science fiction.

Science fiction films have been around for some one hundred years, but science fiction stories were being written long before—in literary history, we can go back at least two hundred years. It may be related to the dawn of the industrial era and its technology, but already in the first half of the nineteenth century, we find novels that in retrospect can be labeled as science fiction. "Gothic novels" like Mary Shelley's *Frankenstein*, as well as Robert Louis Stevenson's *The Strange Case of Dr. Jekyll and Mr. Hyde* are good examples. In both cases, a scientist makes a discovery or builds an invention that pushes the boundaries of what is human possibly, and this is only feasible through that specific discovery. As such, the novels magnify something in our modern existence. It shows the emergence of the overconfident and proud scientist, undeniably in response to people's concerns in the face of ever-progressing technology.

The stories of Jules Verne and H.G. Wells in the late nineteenth, early twentieth centuries mark the beginning of a distinctive literary genre: they consciously make use of the possibilities that are opened up by the tropes employed by Shelley and Stevenson. In their works we find a range of classical themes that characterize much of the later science fiction as well: time travel, space travel, extraterrestrial beings, the future society, altered human capacities, unknown technologies, et cetera.

Science fiction became an established literary genre even before the Second World War and generated many famous authors. Aldous Huxley, George Orwell, and C.S. Lewis wrote their share of science fiction stories, but the genre is defined by later authors like Robert Heinlein, Brian Aldiss, Arthur C. Clark, Isaac Asimov, Kurt Vonnegut, Frank Herbert,

Carl Sagan, and Philip K. Dick. Before the war, two other media were introduced that immediately explored the potential of science fiction: comic books and radio. Comic strips lent themselves to portray the fanciful stories and soon they introduced their own science fiction heroes like Buck Rogers and Flash Gordon. But radio also appealed to the imagination. A famous anecdote relates how Orson Wells produced a radio play of H.G. Wells' *War of the Worlds*, causing major panic among listeners who did not realize it was radio drama rather than real news broadcast. Meanwhile, the genre developed its own following and societies, with their own magazines, meetings, and awards.

As has been said before, science fiction films were first made over one hundred years now, although initially in limited numbers. For obvious reasons: people simply lacked the technical means to recreate the stories in a realistic and convincing way. In 1902, George Méliès fabricated a giant rifle in order to launch a manned missile into the eye of a cardboard moon. With *Metropolis* (1920), Fritz Lang produced a convincing film version of a penetrating futuristic story. It was only after the war that science fiction became an independent genre in film and TV, although it still meant making do with nylon string and extraterrestrial beings looking like dressed-up people. As a result, science fiction long retained the connotation of inferior films, cheap comic strips, and second-rate authors. The 40s were a slack season for science fiction, but the 50s and 60s showed interesting revivals, possibly due to increasing interest in nuclear technology. Even the original TV shows *Dr. Who* (from 1963 onwards) and *Star Trek* (from 1966 onwards) that now belong to the classics only had "cult following" in the early days.

It has been argued that it took until 1968 for science fiction to be taken seriously as a cinematic genre. In that year, Stanley Kubrick's masterpiece *2001: A Space Odyssey* appeared. In this brilliantly constructed story (featuring a spaceship ballet), the intelligent on-board computer HAL, like a kind of Frankenstein's monster, turns against the crew in order to protect them from themselves. Kubrick's film combined high artistic standards, an intriguing plot, and intelligent social criticism in a way that found a response with movie lovers.

Science fiction only gains real popularity among cinema-goers in 1977 with the release of Steven Spielberg's *Close Encounters of the Third Kind* and the first episode of George Lucas's *Star Wars* (although one may wonder if the latter is not fantasy rather than science fiction). Both films appeal to the general public as "popcorn flick," but they also boost technological innovations in the field of "special effects" (like computer-controlled video recordings of models). As a result of these developments, spectacular space scenes could be depicted more realistically, but we are then still in the age of scale-models, background paintings, and adhesive tape.

Nevertheless, from then on, it is impossible to imagine cinema without science fiction. The 80s and 90s saw a steady flow of science fiction movies with increasingly ingenious and award-winning "special effects." This is largely attributable to the development of computer software that mimicked entire scenes as well as actors' movements and facial expressions. Prominent in many of these stories is a feeling of discontentment in relation to the social and ecological consequences of the development of our technological society. Some of them foreground the issue (*Blade Runner*), while others keep it more concealed behind the entertainment (*Terminator*), or behind the cross-over with other genres like horror (the *Alien* series).

Science fiction has seen a string of box office successes over the last years. The top 25 best-selling movies of all time include *Star Wars*, *Jurassic Park*, *Independence Day*, *E.T.*, *The Matrix*, and *Inception*.[4] Technical limitations for filming science fiction stories no longer seem to exist. The age of string puppets and cardboard sets (as in *Thunderbirds* and hundreds of B movies) is over. Lifelike dinosaurs rule the planet, comets are heading towards earth, ice ages and tidal waves imminently rage toward cities; there is no end to it. And in terms of special effects and box office, James Cameron's *Avatar* (2009) topped them all.

The only thing that seems to be missing is an Oscar for best motion picture.

4. Lost in Space: versions, subgenres and family resemblance

The genre of science fiction consists of subgenres, or different "families," with their own characteristics and resemblances. An important adjacent field is that of fantasy. Vivian Sobchack contends that the difference between science fiction and fantasy is that the former wants you to believe or accept something ("it could"), whereas the latter requires the willing suspension of disbelief. We would rather make a case that science fiction revolves around the thought that a certain development in science and technology is extrapolated. Fantasy can be just as speculative at times, but mainly in terms of myths, legends, and fairytales. The distinction, however, is not absolute. We could say that science fiction may be

---

[4] www.imdb.com/boxoffice/alltimegross?region=world-wide (retrieved March 19, 2011).

seen as a form of fantasy, drawing on science and technology as its central plot element.

That being said, a borderline case is a film that is not so much inspired by scientific and technological developments but rather by *nature*: "imagine a world where. . ."; "imagine creatures that. . . or a biotope where. . . ." In some cases, the fairy tale element prevails over the thought experiment, especially when friendly creatures are involved (cf. *Star Wars* and *Star Trek*). In others, the confrontation prevails, mainly when unfriendly creatures are concerned (*Independence Day*). Sometimes we see that the potential of the new world or new creatures not only forms the backdrop to the story (the risk of part 2 and further!) but actually becomes a central issue when something happens that is possible only "there and then." A classic example is *Dune*. Based on a novel by Frank Herbert, *Dune* is about a desert planet whose particular ecology, technology, and mythology bring about events that could not have taken place outside of planet Dune.

When the central elements that comprise science fiction are pushed to the background, some people want to avoid labeling their work as "science fiction" and would rather refer to this derivative as "sci-fi." This term is where we generally find crossovers with other genres: adventure films, disaster films, horror movies, and so on. As has been said, science fiction films do not necessarily take place in space, but there are many space movies—regular hero-and-villain movies, humoristic movies, romantic movies—that happen to be set in outer space. This obviously requires advanced technology, but the plot often lacks serious scientific and technological ingenuity. Sometimes these films contain interesting little philosophies or puzzles, but as long as the plot lines focuses on action, humor, drama, or horror in space, the films cannot be considered real science fiction. Some science fiction films of this kind are explicitly based on classical literature, such as *Treasure Planet* and *Lost in Space*.

It has also been said that science fiction does not need to take place in a different age to be "real" science fiction. At the same time, futuristic movies or films about time travel do form an important subgenre. A number of pessimistic films are set in the future because they speculate that, due to science and technology, we will end up in a degenerate society in years' time. Consider, for example, H.G. Wells's *Time Machine*. (Of course, a few decades in "research and development" need to be bridged to enable time travel.) But there are also science fiction films that are set in the present but still show, to a certain degree, the degenerative nature of humanity, like *Contact, Eternal Sunshine of the Spotless Mind*, and *The Truman Show*.

An important version of science fiction is the type of film that includes a thought experiment on possible developments in *culture*: suppose we descend into barbarism, or dictatorship arises, or a world disaster takes place (such as *The Road* and *The Book of Eli*)? Or, suppose humans can no longer procreate (as depicted in *Children of Men*)? Many of these films contain catastrophes brought about by mankind itself and are oftentimes set in the future (even *1984*). One of the chief examples is the apocalyptic or, more precisely, post-apocalyptic genre, i.e., stories about mankind after a world disaster. *Waterworld* and *The Postman* are good examples of this genre. We could also name this version "dystopian": it refers to a place or time not yet existent (utopia), but one that is, to put it mildly, unappealing. This will be discussed in greater detail in the individual chapters. A classic movie like *Soylent Green* (based on a story by Harry Harrison, "Make Room! Make Room!") also relates to contemporary issues in that it features a "euthanasia clinic." The idea is that the world is overpopulated to such an extent that people who do not voluntarily take their own life, are removed by bulldozers and garbage trucks, and are secretly processed into food for the rest of the population.

5. **General characterization: "The future is our world, the future is our time"**

Meanwhile, the genre's popularity seems to be accompanied by a fascination, an awareness (however indistinct) that science fiction strikes a deeper chord, that it takes us to a level that addresses our culture. People can really fall "under the spell of the Ring," that is to say, they get engrossed in Tolkien's fantasy world. Some people call themselves "Trekkies," while others look down on *Star Trek* because they are *Star Wars* "geeks." It is more than just entertainment: billions are spent in this branch of industry, influencing our culture and therefore ourselves.

When we take this perspective on the central theme of science fiction, we will often find traces of issues that affect the essence of our being. Sometimes they offer a vision of the future that mainly illustrates the one-sidedness of the modern economic, technological, and scientific mentality of our culture. But when science fiction revolves around a scientific and/or technological thought experiment, the possible developments as such are only a part of what makes the genre interesting. The bigger impact is how we as humans would respond to those developments. If science fiction centers on the question: "*what if* certain developments evolve in such a way that things that are not possible today can become real?" then such films eventually deal with something that confronts us with ourselves: what would these developments do with us? What would our attitude or response towards these new possibilities or developments be, and what would they say about us? Something noble? Something perverse? What exactly is in us, and who are we?

In other words, we face questions concerning the nature and meaning of humanity. The thought experiment "what if this or that, what would it do with us?" is ultimately about the question: *What makes us human?* This ancient question is addressed by science fiction in a modern way, by magnifying contemporary issues. In that sense, we can speak of a mirror of our time. We are faced with choices that reveal how we relate to the fundamental certainties of our existence, to the "condition humane." These movies are not simple, meaningless entertainment, but they have an existential dimension. It is no coincidence that ideological and religious themes are interlaced in many science fiction films, especially in *The Matrix*.[5] Various moral and spiritual alternatives are also explored in these films. The critical question for Christians is, then, "*in what form* do these themes and alternatives emerge? And what do Christians have to say about them?" Only then can we speak of an excellent starting point for a Christian philosophy about our contemporary culture.[6]

A useful distinction, derived from Reformational philosophy, is the one between structure and direction.[7] We see various developments and possibilities in our existence, and the direction that they might go is brought to light and is magnified by science fiction. Some cases, such as the excessive control of people by machines, will be critically assessed. We will embrace other directions, because different contexts require different developments. We cannot imagine living without technological aids because they exist and have been an intricate part of our daily life for years. Structures such as technological aids are enfolded in Creation. It is therefore not our responsibility in Reformational philosophy to castigate technological thinking. It is important to identify the structural aspects and normativity of technology and to endorse developments that reveal the structures in responsibility.

Other thinkers have done the same from their perspective: "Knowl-

---

[5] See for example Chris Saey, Greg Garrett, *The Gospel Reloaded: Exploring faith and spirituality in* The Matrix (Colorado Springs: Pinon Press, 2003).

[6] Gregory Bassham, "The Religion of *The Matrix* and the Problem of Pluralism," in Irwin 2002, 111–125.

[7] Richard J. Mouw and Sander Griffioen, *Pluralisms and Horizons*, Grand Rapids: Eerdmans, 1993, 15–17; Albert M. Wolters, *Creation Regained: Biblical basics for a reformational worldview*, 2nd ed. Grand Rapids: Eerdmans, 2005.

edge exists in the world around us," Mark Rowlands says, "and science fiction films provide a vast store of information relevant to the study of philosophy."[8] As a blatantly non-Christian philosopher, Rowlands describes the philosophical theories on topics that are central to science fiction films. He is not the only one. We refer to a series by publishing company Open Court, where able philosophers discuss issues in relation to popular media, from the TV series *The Simpsons* to the rock band U2, but also in connection with science fiction films. Is it instructive as introductory material to philosophy? Absolutely. But, at the same time, it serves as a discussion forum that offers a challenge to Christian thinkers.

6. Conclusion

There is a type of science fiction film where the question "what makes us human?" emerges with particular prominence. This is of importance for the subsequent chapters. There are a number of films where the thought experiment involves the development of a simulacrum of (a part of) humanity. For instance, there are films where "genetic reproductions" of people are produced, with or without certain characteristics (*The Island*, and *Aeon Flux*, for instance). The story usually evolves as follows: the protagonist comes into contact with those simulacra and—due to identity confusion—faces the difficulty of proving himself as human. The plot revolves around the confronting question: "how do we distinguish the truly human from the artificial?" Some of these films are based on the work of science fiction author Philip K. Dick. This theme figures prominently in his books (for example, those adapted to film such as *Blade Runner*, *Impostor*, *Screamers*, and *Total Recall*) and in other areas as well.

How the hero resolves the dilemma between real and fake humanness (*if at all*) obviously depends on the background of the filmmaker. If the film comes from a typical Hollywood studio, the main character will pull through with great valor, gaining the admiration of the heroine. But sometimes we observe a higher goal that the man wishes to achieve or characteristics that make a person truly human: connectedness, selflessness, humility.

Against the background of these existential issues, it is understandable that in this day and age, science fiction films are used alongside other cultural expressions as a medium for what makes us truly human. The main concern in the early stages of the science fiction genre rests in the belief that science and technology lead either to utopia or dystopia. To-

---

8   Rowlands 2003, ix.

day, however, a number of alternative answers to how these developments would affect us as humans actually coexist. Resignation, cynicism, heroism, and desire for a God are related to the quest for the truly human. Below the surface of most contemporary science fiction lies an existential and philosophical level that Christians certainly have something to say about, especially when they are also philosophers.

Chapter 2

WATCHING MOVIES FROM A CHRISTIAN PERSPECTIVE

Neo: "Why do my eyes hurt?"
Morpheus: "You've never used them before."
(*The Matrix*)

1. "Welcome to the desert of the real"

When we walk into our local movie store, we might come across things that are considered questionable from a Christian viewpoint. We know, for example, that the uppermost shelf (or the furthest corner) is filled with movies that we absolutely reject. The films that play a central role in this book also contain much of what we should, in principle, reject: violence, abuse of God's name and so on. Should we visit a place that rents such DVDs? Is there not something fundamentally wrong with the very phenomenon of film itself? Or, should we concern ourselves only with movies that are sold in the local Christian bookstore? Should we restrict ourselves to movies that convey a positive message? Can we simply consider watching movies as a harmless pastime that is useful or pleasant, or can we distinguish between the films that are worthwhile in some respects and those that are not?

Formulated in this way, these questions sound somewhat general but they are more pressing in relation to actual people. For instance, can you justify watching films to your children? If so, what films are fitting? Is *The Matrix* fitting? Such questions are more comprehensive, and they arise again when we take our children to the cinema or rent a movie. Do not even the most decent Hollywood movies already qualify as "worldly"? Should we rent mainly videos that contain a clear Christian message, or has film already reached such a level of familiarity that children absorb it like popular computer games, comic books, and music? Or should we try to show them a range of things that can be appreciated in films from a Christian point of view (and perhaps also the things that cannot)?

These questions do illustrate the heterogeneous relationship between a Christian worldview and the phenomenon of film, omnipresent in contemporary culture. We, the authors of this book, often read and debate about these and other questions with friends, family, colleagues, and students. In books, articles, and web forums, we find lively discussions on those aspects of film that are relevant to our Christian outlook on life. In this book, we wish to contribute to this current debate. We will focus on a number of science fiction films in order to gain a Reformational understanding of certain features of the culture we live in, and its technology in particular. By way of introduction, this chapter situates the topic in a discussion of the relationship between faith and popular media.

2. Why should the devil have all the good movies?[1]

To reiterate, we can, as Christians, have different responses to the phenomenon of film. The differences in attitude are, in part, related to the emphases used in the exegesis of relevant biblical passages as passed on by the various theological and denominational traditions.

William Romanowski divides these traditions into four categories:

- condemnation
- appropriation
- consumption
- distinction.[2]

The first category, condemnation, describes people who may condemn and reject movies in order to keep a safe distance from the influence of the world. Does *Star Wars* contain references to a vague oriental "Force"? Don't watch it. Does *The Matrix Reloaded* show some bare backs? Sinful. In short, films come "from the Devil." Some view alcohol and tobacco as sinful, while others go as far as to say that film and TV are entirely non-Christian.

The second approach involves appropriating the means and practices of the world and using them to preach the gospel. A movie's value is then measured by its testimony: how clear is the Christian message? If a movie does not preach Jesus, or if his name is abused, then the film is unchristian. The *Terminator* and *Alien* movies distort God's creation: don't watch them. But films with "family values" such as *The Sound of Music* are acceptable. In the field of science fiction, you could think of end-times movies such as *Left Behind* as being beneficial. *The Chronicles*

---

1 After a song by the Christian rock singer Larry Norman, "Why should the devil have all the good music?"
2 William D. Romanowski, *Eyes Wide Open: Looking for God in popular culture* (Grand Rapids: Brazos Press, 2001) 12–14.

of Narnia is a good example of a well-received, appropriate fantasy film. By making or watching such films, people wish to distinguish themselves from non-Christian culture by making the voice of the gospel explicit. In many ways, Christian films are like any other in style and structure; however, they profess a more or less overt Christian philosophy.

The third attitude discussed by Romanowski involves Christians who live fully in this world with some sense of propriety but who are in fact hardly conscious of the connection between faith, viewing habits, and film criticism. In actual practice, these individuals consume contemporary culture and its popular media with limited consideration. For some, the underlying reason is the conviction that amusement has nothing to do with spiritual life; others choose to ignore the influence of popular media or are reluctant to gain knowledge of its influence. They see the *Narnia* series, *Harry Potter*, *Contact*, and *Starship Troopers* as the same.

That brings us to the fourth type of response to contemporary culture. According to Romanowski, this approach does not limit the Christian attitude towards popular media to the moral or evangelizing aspects of film, nor does it separate spiritual from cultural life. Christians who adopt this approach towards art and culture choose films that are worth viewing from a Christian perspective while putting aside those that are not in various ways. This approach wants to separate the wheat from the chaff of popular media. It refuses to embrace the Hollywood version of love and romance, of heroism and violence; but, it does appreciate film as a form of artistic and cultural expression. When people are willing to assess films from this perspective, they will not only pay attention to issues of common decency and the preaching of the gospel. Films that offer little in the way of evangelization may still reveal a deep truth about the human condition or give a remarkable insight into God's intention with his creation.

In the science fiction film *Contact* (based on the novel by Carl Sagan), we see the metamorphosis of the main protagonist. From a dedicated scientist who only believes in things that have been scientifically verified, the main character transforms into someone whose own experience cannot be corroborated, at least not convincingly enough for other scientists. At some point she sleeps with a theologian, and, as part of the movie's happy ending, they eventually end up together. There is no mention of Jesus, so the movie might be condemned on both moral and spiritual grounds. But is it therefore a bad or unchristian movie? No. Of course it should not be watched uncritically, but the protagonist's transformation is still a penetrating illustration of a point of debate among

Christians (as well as between Christians and non-Christians) concerning the relationship between faith and reason. Moreover, it paints a compelling picture of the process of conversion. Finally, the subject matter (the ways the possibility of contact with extra-terrestrial intelligence affects the *nobilitas* of the people concerned—whether Christian or not) offers a disillusioning view of the human mind.

3. "Knowing the path and walking the path"

The theological and denominational tradition where Reformational philosophy originates is not part of the culture-shunning approach taken by conservative, reformed Christianity. Rather, it exists alongside this attitude, and the search for conversion and sanctification of the inner self—characteristic for evangelical Christianity—on the other. The tradition, sometimes referred to as "neo-Calvinist," certainly displays marks of cultural criticism and evangelism. But the fundamental issue, as Nicholas Wolterstorff argues, is developing our own opinion of the effects of creation, fall, and redemption.[3] To put it in a nutshell, those who share this reformed or Reformational frame-of-mind regard the creation primarily as God's good gift that comes with reliable conditions, invitations, and instructions for human prosperity. One of the gifts entails that everyone is called to follow God in caring for the creation. In this tradition, cultivating the world before the face of God with kindness, beauty, and wealth is part of the very essence of being human.

Yet people who stand in this reformed or Reformational tradition perceive evil, wickedness, and depravity in human culture as well. They see sin; they see that people go against God's will and that the good possibilities inherent in the creation take the wrong direction. God's goodness does not cease to exist, but the wickedness of mankind pervades every facet of life. Sin corrupts our wills and minds, our thoughts and faith; sin corrupts work as well as family, art as well as technology, rock as well as classical music, films as well as books. Nothing is excluded. Reformed people live in a dynamic relationship with culture, in a dialectic of a critical "yes and no" against "what happens and what is possible," a dialectic also of confirmation and condemnation of human behavior in God's creation.

Thirdly, reformed people—as we dub them—also believe that God reconciles mankind to himself through the sacrifice of Jesus Christ and that he renews us through the Holy Spirit. This, too, has a holistic char-

---

[3] Nicholas P. Wolterstorff, *Keeping Faith: Talks for new faculty at Calvin College*: Occasional Papers from Calvin College (Grand Rapids: Calvin College, 1989) 12–13.

acter in the reformed mindset. God delivers not only the spiritual life of human beings, but the creation as a whole. It is not just us that are in need of sanctification, but our whole existence: society, culture, and nature. We do not only share in God's plan to establish shalom in the created world but also in his plan to establish shalom in the fallen world. In other words, we are called to restore and reform culture through God's Word and Spirit. In short, reformed people see more than just the goodness and the corruption of the creation; they also know themselves to be witnesses of the restoration of the creation. This entails that these Christians will engage in creative activities in the knowledge that we can improve, enrich, or replace art and culture from a Christian perspective. The aim is not to make detached, formal judgments about others, but to shape our "reasonable service" in a way that shows reverence. We honor God by being actively involved in his concern for human existence.

Does this also apply to watching and making films? Absolutely. There was a time when people could object to the rise of the medium of film. But since films—like telephones and computers—are a ubiquitous phenomenon in our western culture and are used for all kinds of purposes, it is just as relevant to ask whether films are Christian as to ask whether phones or computers are Christian. In some respects they are, and in other respects they are not, and sometimes we should act on it, and sometimes we should not. Films mirror culture as a whole and contain specific images that reflect elements of our culture. We are therefore dealing with a Christian assessment of these matters, and that sometimes brings about a transformation of the phenomenon of film and the ways we tackle it.

The holistic nature of the effects of creation, fall, and redemption on our being turns religion (focusing on God or on everything outside him) into the central dynamics of all facets of human existence. We can encounter good and evil, rich and poor, beautiful and ugly, true and untrue, God and devil, in all aspects of film.

4. "Beautiful!"

In the light of such an attitude towards film, there are differences between a *response* to a movie and a *review* of a movie.[4] It is one thing to criticize the content of immorality in a movie, but quite another to evaluate the presence of other qualities, experiences, and beliefs of the people in God's world in the practices and products of the film industry. Such

---

4 Jeffrey Overstreet, Review of *Brokeback Mountain*, www.lookingcloser.org/movie%20reviews/A-G/brokebackmountain.htm, retrieved 10 February 2006.

an evaluation requires detailed observation.

By this we do not mean a more detailed examination of the moral or evangelizing contents of a movie. Undoubtedly, it is possible to further break down those aspects and even provide scales according to which we can judge them. The "Child Care Project Movie Ministry," for example, developed a method that allows films to be awarded with a numerical score of moral suitability. They distinguish different degrees of:

- wanton violence and crime
- impudence and hatred
- sexual immorality
- drugs and alcohol
- murder and suicide
- offense to God.[5]

This is comparable to the evaluation system of movies and pop music by "Focus on the Family."[6] Invariably, they highlight the following points (without adding a quantitative score):

- coarse or blasphemous language
- alcohol and drugs abuse
- sexuality
- violence
- spiritual content and religion
- positive (moral, social) content
- negative (moral, social) content.[7]

Perhaps these criteria also provide a good starting point for those who want to examine movies critically. Yet, along with Romanowski, we can ask if there are no additional qualities where we can critique films. For example, a movie may be about a certain crime and may contain violence and vulgar language; but, in fact, the movie reveals a deep understanding of the human soul in a way that could greatly interest Christians. Roy Anker uses the image of films showing how people's lives catch Light from outside their existence (or not).[8] Science fiction films like *Paycheck* and *Eternal Sunshine of the Spotless Mind* in many respects qualify as contemporary Hollywood movies. But the story about the possibility

---

5   www.capalert.com.
6   These are just two examples. For an overview, see: www.google.com/Top/Society/Religion_and_Spirituality/Christianity/Arts/Movies/Reviews.
7   www.pluggedinonline.com.
8   See Anker's analysis of *The Godfather*, *Chinatown*, and *The Deerhunter*. Roy Anker, *Catching Light: Looking for God in the movies* (Grand Rapids: Eerdmans, 2004).

to erase unwanted parts of the mind also conveys the message that negative memories are indispensible for our happiness, or for our identity. Conversely, it is possible that a film scores high in terms of decorum and faith, while it actually misrepresents human relationships and the way to cope with fate. Evaluating films therefore involves more than merely its moral and spiritual aspects.

Furthermore, we might wonder if films are not supposed to be viewed *as films*—that is, a medium of communication—to begin with. Whether or not a film is morally or spiritually acceptable does not tell us anything about how well it is made, if it makes use of elevating means, or if it adds something to storytelling via other media (such as books). Steve Greidanus draws attention to the artistic and entertaining qualities of film (beside its moral and spiritual value) and also discusses the film's age-appropriateness.[9] In addition, he published several fine articles on his website where he relates films to other films and to literature, providing a broad cultural context for the films discussed.

Jeffrey Overstreet raises still more fundamental questions in his reviews. He points out that a movie may be recommendable even if it contains events that we would reject on moral grounds:

1. Is the film honorable?
2. Is the film artfully made?
3. How effective is the film at what it sets out to do?
4. Is the film worth our time, money, and effort to see it?
5. Did I enjoy it?[10]

Lastly, Romanowski also gives an overview of different perspectives useful for film analysis. He lists the following questions:

1. Cultural product: what does the movie mean to the audience?
2. Cultural landscape: what is considered important in the movie?
3. Artistic features: how is the movie made?
4. Narrative perspective: how is the story told?
5. Reference to the world behind the work of art: what is the mindset, the perspective on issues of religion, God, Creation, mankind, evil, salvation, power, sexuality, violence?
6. Map of reality: how is our existence portrayed and explained?[11]

He points out that these are merely a selection, collected from sources that seemed useful to him in this respect. He classifies them into three

---

9   www.decentfilms.com.
10 www.lookingcloser.org.
11 Romanowski 2001, 156–161.

categories (subject, content, and evaluation), the common factor being the purpose of films: how a film makes visible something essential in our being.[12]

Together, these lists indicate different ways that a film can be considered worthwhile—or even Christian—including the possibility that a film may meet God's intention with mankind in some ways, but not in others. In more than one science fiction movie (*Dark City*, *The Island*), the protagonist finally breaks free of his technology-imposed prison and sees the starry skies, the sun, and the mountains as they really are. An example: At one point in *The Matrix Revolutions*, the protagonists Neo and Trinity soar through the clouds in their hovercraft and, for the first time in her life, Trinity sees the sky, the sun and the moon as they really are. It is a confirmation of the beauty of God's creation when, deeply touched, she exclaims "Beautiful!" Regardless of what else you may think of the movie, it can have "redeeming qualities" that remain invisible when we only look for the moral and evangelizing aspects. This also implies that someone could take an ambivalent critical attitude to a certain movie, or that different religious viewers come to different conclusions. We are, as Wolterstorff states, created as inevitably hermeneutical, searching beings—God can be behind the questions *as well as* behind the answers. He wants us to try to determine his will for our lives.

Meanwhile, it is not always clear why the authors mentioned above chose to discuss these issues rather than others. A certain degree of randomness remains even when their lists are combined. This raises the question of whether it is possible to construct a systematic methodology for reviewing aspects of film. How do we create a broader framework for evaluating, for instance, *The Matrix*? We want to draw on the philosophy of the different aspects of the creation as developed in Reformational philosophy.

---

12   Romanowski 2001, 150. See also his online study guide: www.brazospress.com/Media/MediaManager/eyes_wide_open.pdf.

Chapter 3

REFORMATIONAL PHILOSOPHY RELOADED

Morpheus: "Welcome to the real world."
(*The Matrix*)

1. "You have to see it for yourself"

In the previous chapter we discussed several ways to approach and evaluate films. Movies like *The Matrix*, *The Matrix Reloaded*, and *The Matrix Revolutions* reflect more of God's reality than just the moral and spiritual beliefs that people preach about that reality. In this chapter, we will elaborate on that issue. In what we call Reformational philosophy, a theory has been developed that articulates the diversity of reality.[1] The theory of modal aspects, devised largely by Herman Dooyeweerd and Dirk Vollenhoven, found great response. Their reasons for developing this theory will be dealt with later. For now, it provides us with a "matrix" that can be used for a detailed and systematic discussion of our topic: philosophy and film.

What does this philosophy tell us? It is clear that a single object, such as a movie, has varying aspects. We have already seen that a film has a moral aspect, but it can also have an economic aspect (we pay for it), and a legal aspect (copyright is someone's property). Reformational philosophy does not claim that the things in themselves have different aspects; rather, all things created are subject to various laws and can function within various "spheres of law." Since there are different laws for the functioning of things within reality as whole (such as laws of physics or economic laws), all things have different visible functions or dimensions within these spheres of law. These are also called modalities or modal aspects—that is, the modes of being of things.

Reformational philosophy distinguishes a series of aspects that, to some extent, corresponds with a traditional classification of the sciences.

---

1 See, for example, Roy Clouser, *The Myth of Religious Neutrality* (South Bend: University of Notre Dame Press [sec. ed.] 2005) 243-259.

These are also the disciplines that study the groups of laws where all things operate. As such, we can delineate the numeric aspect, the spatial aspect, the physical aspect, and the biotic aspect. These aspects are irreducible to each other: the numeric differs from the spatial, and the physical differs from the biotic. Life itself, for example, cannot be explained only by using physical laws of matter; life processes are governed by biotic laws.[2] All these aspects follow their own particular set of rules, which explains why one is not more important than the other. The theory of aspects is used as an argument against the different "-isms" in the sciences: economics, the theoretical study of the economic aspect of creation, has to do with its own kind of laws and cannot be reduced to mere logic, power, or survival without loss our economic functioning.

Furthermore, Reformational philosophy claims that the various dimensions of things are ingeniously interconnected. Despite (or because of) their fundamental particularity, a certain order exists among the different aspects. To start with the least complex: it would be impossible to imagine the spatial without the numeric. And, without the spatial, the physical would be impossible. But the opposite is not true: the numeric is possible without the spatial, and the spatial is possible without the physical. The same holds for the physical and biotic: things can exist materially without life, but they cannot live without matter. As such, Reformational philosophy identifies not only a fascinating variety of ways to observe all things in Creation, it also suggests a connection between the different aspects, depending on the laws God established for the world.

There is more to this—especially where humanity is concerned—since a person's functioning reveals even more dimensions. In addition to the aspects mentioned above, Reformational philosophy introduces the psychic aspect, the analytical aspect, the formative aspect,[3] the lingual aspect, the social aspect, the economic aspect, the aesthetic aspect, the juridical aspect, the ethical aspect, and the faith aspect.[4] These universal laws for mankind are also referred to as "norms." To function within these aspects means that people can choose to comply with the prevailing

---

2   Uko Zylstra, "Intelligent-Design Theory: An Argument for Biotic Laws" in *Zygon* (2004) 39.1: 175–191.

3   Dooyeweerd speaks of the historical aspect; to avoid the association with the study of history and the passage of time, some theorists also refer to it as the aspect of formative power.

4   We categorize the psychic aspect (sensation) among the normative aspects, not because we do not agree with Dooyeweerd's hierarchy, but because the perceptive (perception) and the sensitive (experience) are of obvious importance for making, watching and evaluating films.

norms in various ways: one can decide to speak indistinctly, to act unethically, or to squander his resources. It does not alter the fact, however, that it is only by the grace of the existence of those norms that we can assert this. In all his or her actions, in the whole of culture, a person discloses the normative structure of reality. As has been said in the previous chapter, this can take place in different directions, in more or less accordance with God's intention for that sphere of reality.

Much more could be said about Reformational philosophy's "modality theory"; for example, that the different modes of being (to put it philosophically) also mutually refer to each other without losing their distinctive uniqueness. There are indicative analogies that point "forward" to more complex modes of being: for example, in philosophy we find something called the economy of thinking. Then there are analogies that refer "back" to less complex modes: for example, the juridical contains an allusion to the lingual: jurisdiction. Whatever the details, enough has been said in support of the uniqueness of the aspects and the connectedness between them. We can discover all aspects in all things, but in discussing our topic we wish to draw attention to the last set of aspects, the normative ones, in particular. Various aspects have already been mentioned in relation to other authors (see chapter 2), which are correlated to the aspects in the modality theory of Reformational philosophy. These can be summarized as follows:

Dooyeweerd	**Overstreet**	**Romanowski**
Psychis aspect	Did I enjoy the film?	
Analytic aspect		Representation of experiences and reality
Formative aspect	Is the film effective at what is sets out to do?	Cultural product
Lingual aspect		Narrative technique
Social aspect		Vehicle of culture
Economic aspect	Is the film worth our time, money, and effort to see it?	
Aesthetic aspect	Is the film artfully made?	Artistic expression
Juridical aspect		
Ethical aspect	Is the film honorable?	
Faith aspect		Perspective on mindset/religion

2. "I see blonds, reds, brunettes. . ."

What does this mean for the way we watch films? By drawing on a number of examples, we will briefly explain the ten aspects among which we can distinguish several focal points.

Psychic aspect. Overstreet asks: did I enjoy the film? We could also ask, with regard to the content of the film: what emotions does the film evoke? What do I see and what do I experience? Is the film overdone, weak, or constructive in that respect? Romanowski repeatedly points out that Hollywood movies have a preference for melodrama, eliciting a certain emotional response. Another question that is also part of the psychic asks, "Has the film been made in such a way that the different ingredients are plainly visible?" In a documentary on the making of *The Godfather*, director of photography Gordon Willis says he deliberately shot the film to look dark to capture the mood of the story. However, he admits, in some passages it was exaggerated to such an extent that the image itself becomes almost invisible. *The Matrix* also offers numerous examples of moments that produce intense positive and negative experiences. We will confine ourselves here to an observation on the film's loud music. The music is usually energetic, and thereby says something about the harsh circumstances where the characters find themselves: the troublesome reality that exists inside and outside the matrix. There are no embellishments; it evokes a certain experience of our world. What is the effect of this on the psyche of the film's characters? For this reason, *The Matrix* is sometimes referred to as "cyberpunk": it ruthlessly exposes the dark sides of our technological culture.

Analytical aspect. Another aspect of film is that it offers information; it contains a representation of our reality, thereby providing an incomplete picture of our world. We can ask ourselves "what do we learn about the world? What selection of facts, perspectives, and belief systems is being presented? What is being told and what is left unsaid? But also: does it have a logically constructed story? What "plot holes" does the film have? In some films, matters of knowing and thinking as such play a role in the plot. What do the characters know and what do they not know?" A clear theme in *The Matrix* is that some people do not realize they live in a dream world. This is something to reflect on: if someone tells you that all your previous experiences are unreliable, as Morpheus does, can you then trust the person who tells you that all his previous experiences are unreliable?[5]

---

5   David Mutsuo Nixon, "The Matrix Possibility," in William Irwin (ed.), *The Matrix and Philosophy* (Chicago: Open Court, 2002) 35.

Formative aspect. Overstreet asks: is the film successful in what it sets out to do? The content of a film contributes to our culture. We could also ask: how do people respond to the culture *in the film*? What kind of social and cultural mechanisms are depicted? Is the culture in the film constructive, formative of culture, or is it only seeking to produce effect? And as regards the formative aspect of the film's content we can ask "how does the film deal with issues of power, politics, and technology? What direction does it take?" Consider, for example, the post-apocalyptic science fiction films that imagine people after the modern technological culture has collapsed (*Waterworld*, *The Postman*, *Children of Man*, *I am Legend*). *The Matrix* also shows the way that individuals are involved in history. Since machines have taken control and the sky has darkened with black clouds, people do not even know what year it is. But the story has been placed in a cultural-historical context and thereby shows a developmental stage of mankind; it illustrates the cultural aspects that humanity can and cannot influence.

Lingual aspect. In connection to the lingual aspect, we refer to Romanowski's claim that film uses narrative technique or discourse. Plot, script, and dialogue are of particular importance, but so is the use of language: is it coarse, far-fetched, or vulgar, et cetera? In other words, is the film substandard, "over the top," or rich in this respect? Another point of interest is the significance or symbolic meaning of the use certain images, music, clothing, and so on. Much can be said about the lingual aspect in *The Matrix*, especially when looking at the literary allusions, not to mention the biblical ones. Just an example: at the start of *The Matrix*, Neo's computer screen reads: "Follow the white rabbit." He then follows someone who has a tattoo of a white rabbit and eventually he finds Morpheus. Morpheus offers Neo, who probably feels a bit like "Alice sliding down the rabbit-hole," a choice: staying in the virtual dream world or staying in "Wonderland" to discover "how deep the rabbit-hole goes." These quotations clearly allude to Lewis Carroll's *Alice in Wonderland*.[6] Another case in point is when Neo decides to penetrate the matrix, literally and figuratively. Cypher says, "Buckle your seatbelts, Dorothy, 'cause Kansas is going bye-bye." The reference to Frank Baum's *The Wizard of Oz* illustrates that a film can be poor, coarse, or vulgar linguistically, or meaningful, subtle, and harmonious.[7]

---

6   Sarah E. Worth, "The Paradox of Real Response to Neo-Fiction," in Irwin 2002, 184–185;
7   Chris Saey, Greg Garrett, *The Gospel Reloaded: Exploring spirituality and faith in The Matrix* (Colorado Springs: Pinon Press, 2003) 36.

Social aspect. Films almost always focus on human interaction and its social meaning. We are told that people stand in different social relationships to each other, which they can handle in community-sustaining and community-undermining ways. In other words, the social aspect of a film consists in the extent that customs, social conduct, and practices of people advance their coexistence. This usually means focusing on a specific social context: the British army in World War II, a Dutch family in the fifties, an American school in contemporary New York, et cetera. How do people manage authority relations? Do they obey social customs? Many science fiction films concentrate to a great extent on giving "social commentary." Issues of social organization and conduct are not lacking in *The Matrix* either. Morpheus clearly stands in a superior position to his fellow freedom fighters and they treat him with respect. According to his crew member Tank, he was more than a leader, "You were a father to us." It taps into the matter of the significance of male and female roles, which is not unimportant in the defense of Zion.[8] The racial equality aboard the hovercrafts and in Zion is also noteworthy in this respect.

Economic aspect. It is useful to repeat that we are discussing aspects of all things (modes of being), not the things themselves (entities). The same is true for the economic aspect: we do not refer to the economy or even money, because these are things and they function in all modalities. The point is that all things have an economic dimension; we deal with all things in an economic way, too. This is reflected in a film when events and dialogues in the story are inefficient and demonstrate superfluity. It is also reflected when, given the needs and possibilities in the story, choices have to be made. This is true for the filmmaker (does he use too much or too little of something, in such a way that the film suffers from it?), but also within the film itself: the characters struggle with goals, means, and choices. They are unable to do everything at the same time and as a result they meet the viewpoint of stewardship. In *The Matrix Revolutions*, for example, the protagonist Neo is pressed for time when the machines are about to destroy Zion, and he wants to make a pact in order to save humanity. If it was not for the lack of time, he would not be faced with difficult choices or dangerous adventures.

Aesthetic aspect. This aspect compels us to ask questions that most people probably consider obvious in relation to an artistic expression. Is the film cleverly or beautifully made? How are means, techniques, and tropes of the genre employed to represent the story or the world behind

---

8   Cynthia Freeland, "Penetrating Neo: New holes but the same old shit," in Irwin 2002, 205ff.

the work of art? How are the narrative elements developed? What are the acting achievements, the lighting, the recordings, the dialogues, and the music like? Does the film use these elements in a harmonious or in an exaggerated way? In *The Matrix*, it is clear at any rate that special attention has been paid to special effects, costume design, and stage setting. The Kung Fu scenes have been beautifully choreographed. At the same time, the characters' appearance brings along a certain preconception. One of the female protagonists, Trinity, walks around in a tight black cat suit—she is "cool," regardless of how many police officers she liquidates. There is a scene in the first movie where, accompanied by loud, energetic music, Neo and Trinity break into a building and shoot down the guards with tons of automatic guns. The violence is so artfully stylized that you almost forget what it really is—deadly violence. Art is thus not only beautiful; it also depicts different worlds, with different objectives and with more or less "fittingness."[9]

Juridical aspect. The juridical aspect includes issues of legitimacy and justice in the film, such as obedience, crime, and violence. In many contemporary films, the storyline generally hinges on controlled and uncontrolled use of power or violence, or on a tension between legitimacy and lawfulness. Violence can be an illustration of a real situation. Sometimes, however, a film is cunningly crafted in that sense; for example, when the acting is so good, we forget we are sympathizing with a thief, an adulterer, or a fugitive. So in this respect, too, a film can be constructive, overdone, or poor. In *The Matrix*, violence and authority relations are clearly present, in the real world as well as in the virtual world. One could say that Morpheus obeys the prophecy, rather than the "chain of command" of Commander Lock. Rightly or wrongly? The movie also suggests that it is all about a legitimate fight for freedom. Some fight scenes hardly seem to serve a higher purpose than showing people dying a bloody death, others display an aesthetic quality that holds no relevance to other people's lives, and some play a cruel but clear role in the plot.[10]

Ethical aspect. In spite of all, Romanowski does not address the question of ethics separately, and Overstreet confines himself to the question whether or not a film is honorable. Yet, it is obvious that the evaluation of a film is largely of a moral nature. After all, a film also functions in the

---

9   William D. Romanowski, *Eyes Wide Open: Looking for God in popular culture* (Grand Rapids: Brazos Press, 2001) 85–89.

10 Saey and Garrett 2003, 107ff. They address the question whether the fight scenes in *The Matrix* could be seen as a modern idiom for "spiritual warfare" and "spiritual armor."

ethical aspect. In fact, what makes a plot powerful often involves some moral principle that is tested. Good and evil, love and care, justice and injustice are indisputable benchmarks of human existence. Undoubtedly it makes a difference whether or not a film contains references to sex, drugs, materialism, hedonism, et cetera, but a Christian verdict on these things as such does not necessarily do justice to the film as a whole (see previous chapter). In the following chapters, we will explain in greater detail how *The Matrix* offers a palette of ethical possibilities and ideas. A case in point is the moment where Neo is given a choice by the Architect, the creator of the virtual world, to save either Trinity or the population of Zion from dying. But not only such dilemmas are ethical—the attitude to life as embodied by Cypher or by the Merovingian (to name a few bad guys) illustrate that the story itself has ethical aspects.

Faith aspect. The pistic or spiritual aspect can be typified by this quote from the Reformational philosopher Henk van Riessen: "Faith is the function . . . that lifts [individuals] up above [themselves] and [their] situations. It connects him with the future, with the past, and with what surpasses the subject, so that he finds a real or apparently firm ground and final certainty."[11] Films often contain references to God, faith, and religion, or to hope, trust, and a higher reality, i.e., the experience "this is meaningful" or "this I can be sure of." The characters can articulate perspectives on belief systems in the film. The spiritual aspect becomes particularly evident when a character speaks about faith, like Han Solo in *Star Wars* ("It's all a lot of simple tricks and nonsense"). Films can also deal with faith in a superficial way, as in the science fiction film *Serenity* ("It makes no difference what you believe, as long as you sincerely believe it"), not to mention the mere presence of swearing or occultism. *The Matrix* bristles with references to ideological traditions and interpretations of the human condition: in the names of characters (Seraph), phrases ("There is no spoon"), the use of symbols and metaphors (Neo is reborn), and events (counselor Haman gives an "opening prayer").[12] The unwavering faith of Morpheus in the prophecy, and Neo's struggle with the significance of his choices, clearly demonstrate the spiritual aspect of the film. We will return to this issue in later chapters. When we state that every film also has a spiritual aspect, we refer mainly to the visible manifestations (whether positive or negative) of spirituality in the film, rather than the outlook on life that drives the film maker. This question

---

11   Hendrik Van Riessen, *Wijsbegeerte* (Philosophy) (Kok: Kampen, 1970) 188–189. Own translation.
12   Saey and Garrett 2003.

will be discussed separately below.

3. "It is inevitable"

There is another element in Reformational philosophy that may be useful in the discussion of films. The objective of the theory of aspects in Reformational philosophy can be explained in such a way that it accounts for the possibility and nature of theoretical or scientific knowledge. And, more importantly, it can provide an argument in support of the idea that the achievement of knowledge is difficult to separate from *worldview*. Because this line of thought contains similarities with the making of films, we will briefly summarize the argument here.

We have already seen that over a dozen aspects can be distinguished in all things. Everything functions within all spheres of law. The first step in the argument is that humans and human perception (consciousness) actually function in all these respects, too. Seen in the light of the theory of aspects in Reformational philosophy, a person can thus direct his or her consciousness toward the world in different ways: aesthetic pleasure, historical consciousness, sense of justice, moral intuition, religious faith, et cetera. The second step in the argument is that, besides the aforesaid modes of experience, a person can also analytically examine the laws of phenomena. When that happens, he can gather knowledge and articulate his findings in a way that we call "theoretical." Scientific practice is possible because the analytical function of human consciousness is consciously directed at one of the modal functions and laws of one or another object of study.

But—so the argument continues—it is always a human being, a concrete person, gaining scientific knowledge. This entails that the knowledge the scientist acquires about reality is connected within the whole of reality to the way the researcher views his topic, the laws concerned, and himself. That is the third step in the argument. Regardless of the method used, the researcher can only gain knowledge against the background of his or her beliefs about the order of things.[13] The acquisition of theoretical knowledge is inevitably supported and limited by the way the theorist approaches the "origin and purpose and meaning" of reality. In other words, his religious orientation works as an "archimedean point" for his theorizing. Dooyeweerd refers to "ground motives" as the root from which the branches grow on the tree of theoretical thought.

Making, watching, and discussing films differs from gathering theo-

---

13   Cf. Nicholas Wolterstorff, *Reason within the Bounds of Religion* (Grand Rapids: Eerdmans, 1984).

retical knowledge. One can wonder whether the acquisition of knowledge does indeed take place in the way described above. Still, this line of reasoning is of interest to our present topic. In the wide world of film, too, analytical skills are activated and, although a knowledgeable critique of film may not exactly be the same as science, the increasing number of studies on film shows that there is at least a basic similarity between them.

The point, however, is that the picture of knowledge acquisition presented above shows much resemblance with the work of a filmmaker. In fact, a filmmaker quite literally does what happens *in abstracto* in theory formation, namely the putting together of perceptions and events, the process of giving *meaning* to image and sound. In the same way, the filmmaker's decision to arrange certain images in a film cannot be separated from who the filmmaker is as a person. His worldview has an influence on his representation of reality and this materializes in his film and oeuvre.

Romanowski adds an important comment: we should be careful not to over-individualize the issue.[14] After all, every filmmaker is part of a team, a company: a creative community. Making films is a wonderful example of performing social practice in the sense of a collective, socially and historically rooted activity with its own dynamics, rules, aims, and means.[15] No filmmaker, not even the most individualistic one, is an isolated being. Films reflect elements of cultural and philosophical "movements," whether the film maker uses them intentionally or unintentionally.

It does not alter the fact that, according to Reformational philosophy, performing such practice cannot be separated from the practitioners' mind set, from their take on the practice, or the ultimate goal or purpose of the practice itself. This is certainly true for a film that is written by the directors themselves. It is not for nothing that Romanowski asserts that a film is a message, a text that contains a certain interpretation of reality; composing and explaining it can never be detached from the "I" positioned in that reality. It makes a difference, for example, which producer or studio the filmmaker chooses. Usually an ethos or ground motive can be detected in a film or oeuvre that helps to explain the way in which the different aspects are manifested. A film that appears adequate in several respects can still reveal a cynical perspective on life; and, a chaotic, vio-

---

14   Romanowski 2001, 63.
15   Alasdair MacIntyre, *After Virtue: A study in moral theory* (London: Butterworth, 1983) 183.

lent film can eventually display a loving view on God, humanity, and the world.

In short, what Dooyeweerd terms a "transcendental critique of theoretical thought"—the notion that the act of theoretical thinking can never be separated from the pre-given relation of the thinking actor to reality—suggests that making, watching, and discussing films cannot be seen as existing independently of the religious attitude of the person filming, watching, and reviewing. Our world exists in a structured manner, and humans are structured subjects who respond to it in a structured manner. That is why we will also try to obtain anthropological insights about the film *The Matrix*: what can be said about the ethos in the film, about people's stance toward life?

4. Neo *Respondens*[16]

We do not intend to present a checklist to help people screen films or eligibility criteria for a yet-to-be funded award from the Reformational Academy of Film and Philosophy (and the "Herman" goes to. . . ?). Our aim is to offer an accurate tool for watching and discussing films so that the effect of the goodness of God's creation, the depravity of fallen humanity, the joy of salvation through Jesus Christ, and dimensions of Holy Spirited renewal can be recognized and used in service to God and humanity.

There is no set script for this; it is important that everyone acts out of his or her own responsibility. It is not everyone's obligation to always think through every film. We fulfill different roles in the practices we live, and how we deal with the phenomenon of film depends on the tuning of our responsibilities and roles. My role as teacher, father, journalist, or minister possibly involves a responsibility to know what kind of films young people might watch. In that case, I will need an instrument to discuss different aspects of films with these teenagers. Especially if I know, as we saw in the previous section, that through making and watching films, an atmosphere may be created that is not necessarily inspired by the Holy Spirit. Others who have entirely different responsibilities and may want to only enjoy films could, in most cases, trust other people's judgments. Since I cannot assess the quality of the work done by a policeman, a chemist, or a laboratory assistant, I trust those who have discernment in these matters. In other matters, however, I do want to know what's what, and I therefore have the responsibility to use my own eyes.

---

16 After Hendrik Geertsema's phrase in "Homo Respondens: On the historical nature of human reason," *Philosophia Reformata* 58 (1993): 120–152.

Chapter 4

THE MATRIX: A MODERN MYTH?

Obi-Wan Kenobi: "You've just taken a step into a larger world."
(*Star Wars: A New Hope*)

1. The Hollywood Code

In the previous chapters we have referred to some of the most remarkable science fiction films of recent years, and we will continue to do so more often in the following chapters. It is useful to be familiar with the context of the scenes from *The Matrix* that will be used to illustrate our argument. This chapter therefore maps out the storyline of the trilogy in greater detail. We will not take a chronological approach, however, which would be slightly haphazard and give away too much of the story. Instead, we will draw on Joseph Campbell's model of storytelling, based on his comparative studies of folktales, legends, myths, and sagas from numerous cultures.[1] This model has been used in connection with epic movies in the vein of *The Matrix*, including some superhero movies and *Star Wars*.[2] It is sometimes even used by filmmakers to structure stories in a way that makes sense to the audience. For that reason, it is sometimes referred to as the Hollywood code.[3]

Campbell brought to light a common structure (the so-called "monomyth") in the literary and religious stories that have been told in cultures throughout centuries. These myths and stories reveal something about how people relate to existence and its fundamental characteristics; for example, the origin of the cosmos, the significance of evil, the relationship between the masculine and the feminine, the heart and goal of being human, and so on. In these stories, Campbell asserts, we can discover a tripartite structure of a hero who (1) has to leave behind his everyday life, (2) to undergo a crucial experience, and (3) must return to

---

1   Joseph Campbell, *The Hero with a Thousand Faces* (Princeton: Princeton University Press, 1949).
2   www.moongadget.com/origins/myth.html (retrieved March 19, 2011).
3   Laurent Bouzereau, *Star Wars: The annotated screenplays* (New York: Del Rey, 1997) 35.

his life in a new light. To correctly understand this structure, we will first explain it by means of another film that is fit for the purpose, and then apply it to *The Matrix*. Doing so also makes for a good exercise to try and recognize this structure in other films.

The original *Star Wars* trilogy provides us with a clear example of Campbell's model. This trilogy covers Episodes IV to VI of George Lucas's famous space saga, released between 1977 and 1983. (Episodes I to III take place earlier in the timeline but were produced later.) In these films, the protagonist, Luke Skywalker, (1) is driven from his simple farmer's life, to (2) play a role in the battle against his vicious enemy Darth Vader and Vader's master the Emperor, and so (3) restore peace and justice in the galaxy once again.

According to Campbell, each of these three stages can be subdivided into different steps. At the

> **Campbell's monomyth**
>
> **I: Departure**
> The call to adventure
> Refusal of the call
> Supernatural aid
> Crossing the first threshold
> The belly of the whale
>
> **II: Initiation**
> The road of trials
> The meeting with the goddess
> Temptation away from the true path
> Atonement with the Father
> Apotheosis (becoming god-like)
> The ultimate boon
>
> **III: Return**
> Refusal of the return
> The magic flight
> Rescue from without
> Crossing the return threshold
> Master of the two worlds
> Freedom to live

beginning of the adventure, the hero initially receives a call to action. He is summoned to leave his everyday world behind and embark on an adventure. In the old *Star Wars* trilogy, for example, Luke finds the hologram of Princess Leia: "Help me Obi-Wan Kenobi; you're my only hope." Initially, the hero refuses the call to adventure. Luke, nephew of farmers, claims he has to stay for another season to help with the harvest. But soon the hero leaves as a result of external aid or impetus. In *Star Wars*, Luke is given a helping hand by his mentor Obi-Wan Kenobi, who is endowed with special powers and wisdom. Circumstances arise such that Luke is forced to escape with him. The hero thereby crosses the threshold of his old existence, leaving it behind to face the unknown. Generally, the next step involves the hero encountering serious difficulties. He might become entangled in the web of the enemy through imprisonment, or at least ends up in a situation that by no means meets his expectations. Campbell describes this, with a clear biblical reference, as "the belly of the whale."

The second stage of the adventure—the initiation into the new "being"—is marked by a road of trials, sacrifices, and tests. To become a Jedi knight, Luke is trained by Obi-Wan Kenobi and the old, wise Yoda so he can defeat the enemy. Typically, the hero will then meet a motherly figure or female counterpart who helps him to grow into the person he is meant to be. Sometimes—but not always—she is a goddess or a sexual partner. In *Star Wars*, Luke serves Princess Leia, who ultimately turns out to be his sister. The different temptations to leave the true path and to give up the true goal are also part of this stage. Luke is impulsive and would rather help his friends than complete his training. He is tempted to follow hate and anger instead of composure and self-control. Yet, he perseveres.

As a rule, the second stage of adventure contains a confrontation with the male or fatherly figure so the hero can become the warrior or leader he is meant to be. Luke discovers that his opponent, Darth Vader, is actually his biological father, Anakin Skywalker. This means he will have to defeat his father or be reconciled with him. In the next step, Luke does both: he conquers Vader but does not kill him. He even surrenders to him, thereby evoking in Vader whatever goodness is still left in him so as to prevent him from destroying Luke. This is the apotheosis, when the hero becomes who he was destined to become: he finds the treasure, sees the light, or wins the victory. Because of his victory and his self-sacrifice, Luke—or through him, Vader/Anakin—becomes a Jedi knight, restoring peace and justice in the galaxy. This final outcome also entails the return of the hero: he will leave the world of adventure behind and live in a new light. In the *Star Wars* trilogy, this stage is marked by festivities in space including all the figures we have encountered in the films.

But before we get to that point, Campbell describes the third stage of the adventure: the process of returning. Usually, the hero initially refuses to return. For example, when the dying Darth Vader urges Luke to flee the exploding space station, he initially tries to save his father rather than his own skin. This dilemma is followed by what Campbell terms "a magic flight." The hero manages to escape the world of trials just in time. Luke carries his father's body with him, just before the space station collapses. In most cases, this again requires external aid. In the final part of *Star Wars*, a space war creates great confusion when the unexpected return of a friend enables Luke to flee the nest of enemies. In that way, the hero again crosses the threshold between the world of adventure and his old familiar world. At the end of the third film, *Return of the Jedi*, Luke is clearly back in safe territory, back in the reality of everyday needs and habits like eating and drinking, laughter and music. Two things are evident in the final stage of the adventure: first, the hero is acknowledged in both worlds as someone of quality, a master or ruler; second, the hero has brought about a liberation that allows life to be lived without threatening restrictions. At the end of *Star Wars*, Luke has restored justice and peace in the galaxy, or, as the viewers know, re-established the balance between the light side and the dark side of the Force that surrounds us, penetrates us, and binds the galaxy together.[4]

2. The Matrix: "Do you think you are the One?"

Concluding our brief survey of Campbell's theory of the monomyth, we immediately encounter interpretative problems (aside from the cogency of Campbell's theory as such). On one fan site, Campbell's schema is applied exclusively to the oldest *Star Wars* film, *A New Hope*,[5] while we used examples from the entire old *Star Wars* trilogy. It has now been succeeded by a new trilogy, where Luke Skywalker's father Anakin Skywalker—who later becomes Darth Vader—figures as the protagonist. A prophecy about him predicts that he will restore the balance of dark and light, or justice and peace. It is clear that the model needs to be filled out in a different way (and, even still, several interpretative problems continue to exist; for example, whether Campbell's elements need to occur exactly in this order).

A similar notion applies to *The Matrix*. Campbell's model is also used on the aforesaid website to reconstruct the first *Matrix* film.[6] Our focus here, however, is the trilogy as a whole. Keeping in mind then that there are different ways of representing the plot—even within the same model—we will present the monomythical structure of the storyline of the *Matrix* trilogy.

Roughly speaking, the first *Matrix* film consists of three parts marked by moments where the main characters enter or leave the matrix, a virtual world. At the beginning of the film, we seem to fall right into the middle of the story. Only later do the pieces fall into place; indeed, the film's structure is not instantly clear to those watching it for the first time. It starts with two main characters, Trinity and Morpheus, who are looking for one Thomas Anderson, alias Neo, along with several "Agents."

---

4   Bouzereau 1997, 59.
5   www.moongadget.com/origins/myth.html (retrieved March 19, 2011).
6   Chris Saey and Greg Garrett, 28–29, 71–80, and 159–160.

When they find him, Morpheus tells him that the world where all this happens, apparently 1999, is actually a dream world, the matrix, from which they can free him. This is the famous scene where Morpheus offers Neo the red and the blue pill: if he takes the blue one, he will stay in the matrix, if he takes the red one, he will be disconnected from it. When he chooses the latter, Neo is collected by Morpheus' ship, the Nebuchadnezzar. We are introduced to his crew, including Cypher, a cynic; Tank, the fearless operator; and Mouse, a somewhat immature figure.

The events aboard the ship make up the middle part of the film. Morpheus turns out to be "a morphing Orpheus, a black White Rabbit, an R-and-B Obi-Wan Kenobi, a big bad John the Baptist, a Gandalf who grooves: every wise guide from literature, religion, movies and comics."[7] He explains that the dream world Neo lives in is actually a system maintained by intelligent machines, supercomputers. In reality, it is closer to 2199. Humans are artificially kept alive by the machines that use their body heat as energy source. People are thus unwittingly controlled by a computer program that gives them a collective mental projection of reality, known as "the matrix." In reality, their bodies are kept in vats filled with a preservative fluid. The name "matrix" is brilliantly chosen in this regard, since the Latin word means "mold" as well as "womb." Only a few people have been freed from this state, and they are trying to free others as well. Morpheus, the leader of the rebels, believes in the Oracle's prophecy that someone will come to liberate the human race from the machines. He believes that Neo is this savior, the One. He trains him so he can manipulate the software of the matrix.

In the final part of the film, the crew enters the virtual world of the matrix to consult the Oracle as to how they can help Neo realize that he is "The One." He does not seem convinced. On the way back, they are attacked by the Agents we saw earlier in the film, who are actually powerful, autonomously cognitive programs responsible for guaranteeing the continued existence of the matrix. It turns out that Cypher has betrayed them; Morpheus is captured by their pursuers. Only then does Neo start to believe that he has special powers to manipulate the matrix in such a way that he can defeat the Agents, headed by Agent Smith, and save Morpheus.

Neo's first resistance to leaving the matrix arguably reflects Campbell's notion of the "call to adventure." Early on, he is summoned to leave his old world, as seen in the choice between the red and blue pill. But the appeal to pursue a higher goal and to give up several things in the

---

7 Ray Corliss, quoted in Saey and Garrett 2003, 27.

process only becomes clear as the story develops: he must become "The One," a savior. Neo refuses to accept this: "the refusal of the call." He does receive, in Campbell's terms, "supernatural aid" from the Oracle, but it remains ambiguous. He thinks that the Oracle is telling him that he is not the savior. Even rather late in the film, he says to Trinity: "I'm not the One, Trinity. I'm sorry, I'm not. I'm just another guy." He does believe at that time that he has the power to rescue Morpheus from the Agents and to defeat them. This discovery is what actually brings him to what Campbell calls "crossing the first threshold": he embarks the world of his new destiny. When that happens, the viewers already know that Neo is the One. Morpheus, the prototypical "old and wise helper," makes the point that Neo's identity is revealed not through what Neo believes, but through what he does. As a result of his actions, Neo ends up in "the belly of the whale" (Campbell), the anything-but-glorious reality of the Nebuchadnezzar, seeking shelter in the old sewers of the former human world, where all food tastes the same. At this point the second movie starts, with Neo sighing despite his newfound status: "I wish I knew what I am supposed to do."

Excursus: *The Animatrix*

How did it come to this? What has changed between the end of the 20th and the end of the 22nd century? *The Matrix* only tells the story bit by bit. The makers of *The Matrix*, the brothers Andy and Larry Wachowski, discuss this history in greater detail in two animated short films on the DVD *The Animatrix* under the title "The Second Renaissance." In the early 21st century, people created machines with artificial intelligence (AI). These machines were not only endowed with a brain, they also developed consciousness. The machines' reflection on their own situation led to a struggle for their rights: they wanted to be treated as free and equal citizens. But the humans refused to meet their demands; in fact, they banned machines from society altogether. The machines then founded their own city and named it 01; a reference to the binary code that enabled consciousness. Ironically, according to "The Second Renaissance," the city is situated somewhere between the Euphrates and the Tigris, which was also, according to the Bible, the birthplace of civilization: the Garden of Eden. The founding of the city 01 brought about a war between humans and machines. In a desperate attack against the machines, which were dependent on solar power, the humans filled the sky with thick black smoke, but without success. In the end, the humans were completely destroyed and their bodies were connected to the matrix to be used as a new energy source. A few survivors escaped to the underground sewers where they founded the city of Zion.

3. The Matrix Reloaded: The One, or Neo 6.0?

The second installment of the trilogy, *The Matrix Reloaded*, can also be subdivided into three parts, the distinction again largely determined by action in and outside the matrix. Early on in the film, we are introduced to the world of Zion, the last safe refuge of humans outside the matrix—this is called "the navel of the world" in Campbell's monomyth. We discover how the remnants of humanity survived as they escaped the machine world. An important element of the plot, provided at the start, is that Sentinels—hunting machines—are heading towards Zion to destroy it. In order to know how to proceed, Morpheus and his men are waiting for a message from the Oracle. When they receive a message from her, the story shifts to the world of the matrix where the second part of the film is set.

This is evidently the start of Neo's "road of trials"; he no longer resists his status as "The One." But what does that imply? In the matrix, the Oracle instructs Neo to find the Source, where all programs controlling the matrix (and thereby humanity) are made, and all programs that threaten the matrix (and thereby liberate humanity) are erased. To access the Source, Neo first needs to find the Keymaker, a program controlled by another program called the Merovingian. The latter represents the system; he only believes in cause and effect and out of self-preservation needs to prevent the Keymaker from handing over the key to the Source. The central part of the movie focuses to a large extent on how Neo, Morpheus and Trinity battle with some of Merovingian's accomplices. It also turns out that the old opponent, Agent Smith, has found a way to break away from his original task in the matrix, to replicate himself, and to take over the personality of a certain Bane, even outside the matrix. He has become a loose cannon that can lead to the destruction of humanity and machine.

Neo is able to find the Keymaker with the help of Merovingian's partner Persephone. We may safely say that this embodies Neo's "meeting with the goddess." Persephone undeniably represents the Feminine that determines the goal of Neo's adventure. She asks him to kiss her the way he kisses Trinity, and only then will she release the Keymaker. Bringing this sacrifice gives Neo access to the Source.

Thematically speaking, the third part of the film ties in with Neo. When he reaches the Source, Neo has to position himself in relation to a father figure, the Architect, the creator of the matrix. After all the action, the plot deepens again in this encounter. The Architect tells Neo he is actually the sixth version of a program designed to guarantee the existence of the matrix. The imperfect brains of the human bodies connected to the matrix reject software of mathematical perfection. Human brains only accept a matrix that offers them (the illusion of) choice (although a small number reject this too and take refuge in Zion). By the time this threatens to make the matrix uncontrollable, the character of "The One" is activated in the software. By using a certain code he will cause the matrix and its human batteries to go extinct. The One will start a new generation of humans and a new version of the matrix will be "rolled out."

Our protagonist, however, does not display a general preference for ensuring the existence of the human race but rather a personal love for an individual—also in response to her love for him. In the meantime, Trinity has entered the matrix to save Neo, but she runs into Agents who are about to annihilate her. The Architect offers Neo the choice to not enter the code and return to the matrix to rescue Trinity, or to enter the Source and save humanity. In Campbell's terms, Neo must reconcile with the Father figure in this scene. He rejects the possibility to act as a mere function of the matrix and chooses to save Trinity from death.

The way that the second *Matrix* film ends also deepens the plot in a surprising manner. Outside the matrix, the Nebuchadnezzar comes under attack by Sentinels and the crew is forced to abandon the ship. During the pursuit, Neo discovers that he not only has the power to manipulate the software within the matrix, but he now also has the ability to "sense" the machines outside the matrix. He is capable of stopping them in the same way he can stop bullets and other things in the matrix. His ability to do extraordinary things in the matrix already impressed people, but now it appears he has supernatural powers outside the matrix that are unrelated to his function as "The One." This is what Campbell calls the apotheosis, or "becoming god-like"; he obtains the highest power, the mythological elixir of life. He collapses due to the exertion, and the film ends with a cliffhanger as he lies in the sickbay with Bane.

4.   The Matrix Revolutions: "Why, Mister Anderson, why?"

The start of the third and last *Matrix* film ties in closely with this. Neo appears to be mentally present in the matrix, without being physically connected to it. That is to say, he finds himself somewhere between the matrix world and the machine world. Other travelers in that place are the characters Trainman and the little girl named Sati and her parents Rama-Kandra and Kamala, who will be discussed in later chapters. The first part of the film takes place in the matrix where Trinity and Morpheus are in search of Neo. In order to bring him back, they will have to appeal to the Merovingian for assistance. After they have accomplished this, Neo has to visit the Oracle one last time. He will have to find a way to prevent the machines from winning the war against the humans. In addition, Smith has clearly evolved into his ultimate opponent.

Once outside the matrix, the magnitude of the threat to Zion posed by the Sentinels becomes clear. It is decided that Morpheus has to return to Zion while Neo, accompanied by Trinity, leaves for Machine City to find a way out of the war. A massive war for Zion also breaks out in full force. Bane, the crew member whose personality has been "infected" by Smith, turns out to be on board of the ship. A fight follows, where Neo loses his sight, but they manage to kill Bane and reach Machine City.

Neo confronts the machines with the fact that the Smith program also threatens the existence of the machines; he proposes that their war against the humans end if he succeeds in defeating Smith. In Campbell's terms, this is the hero's "ultimate boon," the highest good: peace. The next part of the film is set in the matrix, where the ultimate fight between the savior and his last enemy is carried out. Eventually, Neo defeats Smith by sacrificing himself and assimilating Smith: by absorbing all his destructive power, he destroys both himself and Smith, and so wins. This means that Neo refuses his own return to the matrix and to Zion (Campbell: "refusal of the return"). Now that Neo and Smith are gone, a new existence can be built for humans and machines, and the machines end their destruction of Zion. Morpheus and the others realize that Neo has managed to save humanity: mythologically speaking, he is acknowledged as the "master of the two worlds" and has accomplished "the freedom to live." His body is respectfully carried away to the light by the machines. We could thus with Campbell speak of a "magic flight," to heaven if you like. Neo is carried away as if on a crucifix. Because of Neo's self-sacrifice, however, there is no real "rescue from without" or "crossing the return threshold" as is generally the case in the hero's journey. This is where the comparison with Campbell's model ends. Nevertheless, Neo has achieved the goal of his quest and has achieved greatness in both worlds.

One of our film friends once suggested that the three *Matrix* films have as themes faith, love, and hope respectively. Perhaps this is true. The question, at any rate, is what a modern mythological mix as this tries to tell us. If the archetypal story elements of Campbell's monomyth seek to touch a chord in our contemporary soul, what do they tie in with? What are the "Big Ideas" and "Big Questions" these films grapple with? We will tap into this in the next chapter.

Chapter 5

THE LION, THE WITCH, AND THE MATRIX

> Morpheus: "The matrix is everywhere.
> It's all around us, here even in this room.
> You can see it out your window, or on your television.
> You feel it when you go to work, or go to church or pay your taxes.
> It is the world that has been pulled over your eyes
> to blind you from the truth."
> (*The Matrix*)

> Cypher: "Buckle your seatbelts, Dorothy,
> 'cause Kansas is going bye-bye."
> (*The Matrix*)

1. "Do you want to know what it is?"

In addition to the overview of the storyline of the *Matrix* films presented in the previous chapter, it is also useful to introduce some of the central features and motifs of the story. This includes addressing more general themes that will be expanded upon later.

After the opening scene in *The Matrix*, the first word that appears on screen is "Searching. . ."—a query on Thomas Anderson's computer. The way that the film presents the story to the viewer reflects something of what can be called "the postmodern mindset." This refers to the awareness of many people that old familiar truths and certainties have been lost in our time. The view of a person's place in the cosmos and his or her attitude towards life has changed fundamentally. Scientific insights and technological progress are undoubtedly accountable for this (a topic we will address in subsequent chapters). Not only do we perceive our world differently, but we also create our own reality, our own lifestyle, and our own bodies. We can move to the other side of the world within a few hours or even with a simple mouse click. We find ourselves in an infinitely vast universe, characteristics of which we were previously unaware.

At the same time, we are told that human beings have developed over millions of years as a result of a chain of coincidences. The history of humanity does not contain one single truth but can be approached from different perspectives. What we consider to be good and true does not necessarily correspond with the views of people in other parts of the world. The familiar interpretative frameworks of western thought lost credibility in the eyes of many, or they appear to be subject to change. Grand Narratives have lost their persuasiveness. Some even suggest that there is nothing outside our interpretative frameworks to which they might refer. In short, people are in search of something; they feel as though they are falling, without a safety net. (Is that why *The Matrix* often has people leaping from the dizzying heights of skyscrapers, or falling down from them?[1])

This postmodern mindset ties in with our anxiety about the extent to which our experiences and interpretations of the world will prove to be correct. Is there any possibility at all to come into contact with the reality outside us, or has this contract, in George Steiner's words, been broken? Do our interpretative frameworks perhaps only mutually refer to each other? An even more pressing question: what is still genuinely, authentically human, worth living for? How can we establish that? Has it not, at best, become makeable, and thereby arbitrary? What is still true and untrue, good and bad, healthy and sick, sacred and sacrilegious, useful and useless? Where is the distinction between illusion and reality, and who or what helps us escape from that dilemma (if at all)? The mood of our time resembles that of the Book of Ecclesiastes: is everything vanity and a chasing after wind? This is the theme of *The Matrix* trilogy as a whole. The most literal allusion is probably the moment when Neo is seen holding a copy of *Simulacra and Simulation*, a book on illusion and reality by the postmodern philosopher Jean Baudrillard (a little filmmakers' joke: Neo's copy is a hollow fake).

Some artists respond ironically to the matter and turn human history into a search for a, perhaps imaginary, conspiracy. Books like *Foucault's Pendulum* by Umberto Eco and *The Da Vinci Code* by Dan Brown fall into this category. It also includes films like *National Treasure* and, in the area of science fiction, the *X-Files*. An alternative response to the postmodern condition is a longing for truth, beauty, and goodness and a belief in an order that will restore our entire world of experience. Examples include *The Lord of the Rings* by J.R.R. Tolkien, the *Chronicles of*

---

1   Chris Saey, Greg Garrett, *The Gospel Reloaded: Exploring spirituality and faith in* The Matrix (Colorado Springs: Pinon Press, 2003) 63.

Narnia by C.S. Lewis and, although probably closer to fantasy than to science fiction, the *Star Wars* saga by George Lucas.

It is no coincidence that a whole string of films brings this mindset to light, usually through a thought experiment that contrasts something artificial with the question of what is truly human. Imagine we could manipulate memory; what parts of it would prove indispensable (*Eternal Sunshine of the Spotless Mind, The Final Cut, Total Recall, Paycheck*)? Imagine we could manipulate our lifespan; what would make it worthwhile (*Blade Runner, Aeon Flux*)? Imagine we could produce a simulacrum or clone of a complete person; what would be the distinguishing aspect in our identity, and what significance would corporality have in this respect (*Impostor, The 6th Day*)? Imagine we could predict future events; what would it say about freedom and guilt (*Paycheck, Minority Report*)? Imagine we could create intelligent machines or robots; how would it affect human reason, emotion, et cetera (*Artificial Intelligence, 2001: A Space Odyssey, I Robot, Bicentennial Man*)? Imagine humans can no longer procreate; what would give hope and meaning to life (*Children of Men*)? The argument is clear: the postmodern condition confronts us with a great number of questions concerning the fundamental characteristics of human existence that used to be familiar and self-evident. But their authenticity—or the lack thereof—is now being magnified and tested in films (and, of course, not only there).

We find this same anxiety in Thomas Anderson, the protagonist of *The Matrix*, in the beginning of the first film. He is searching the Internet for an answer to his question whether there is something "wrong" with the world, or if there is something different, something more. He wonders to what extent something or someone could tell him if a veil of experiences and meanings has been pulled over his eyes to make him believe in a reality that is not actually real. He receives a sign from the outer world to wake from his postmodern nightmare: the words "Wake up!" appear on his screen. After following the instructions, he meets Trinity, who confirms his suspicions:

> Trinity: "I know why you're here, Neo. I know what you've been doing. I know why you hardly sleep, why you live alone, and why night after night you sit at your computer: you're looking for him. I know, because I was once looking for the same thing. And when I found him he told me I wasn't really looking for him; I was looking for an answer. It's the question that drives us, Neo. It's the question that brought you here. You know the question just as I did."

Neo: "What is the matrix?"

Trinity: "The answer is out there, Neo. It's looking for you and it will find you if you want it to."

It has been asserted in an article in *Wired* magazine that science fiction writer Philip K. Dick turned metaphysics into a "whodunit"[2] (consider, for example, the success of *Paycheck, Minority Report, Blade Runner, Impostor,* and *Total Recall*). Similarly, it could be said of the makers of *The Matrix* that they turned the postmodern nightmare into a science fiction or Kung Fu film.[3] Clearly, the film not only contains existential markers but also philosophical ones: epistemological (what can you know, what is experience), metaphysical (what is it to exist, what is freedom), ethical (what is good, what is love), techno-philosophical (what is the nature and boundary of technology, what is the difference between humankind and machine), anthropological (what is free will, what are emotions). Some of these issues will be explored in greater depth in this book.

2. "There is no spoon"

Before we embark on this exploration, we wish to add two preliminary remarks. In the first place, *The Matrix* also centers on the question whether there might be a hidden reality behind our reality and if it affects our existence. This is not merely an image that can be found in other stories such as *The Wizard of Oz, Peter Pan, The Chronicles of Narnia,* and *Alice in Wonderland*. In these stories, the main characters can acquire qualities or insights they need in the real world. Instead, we are dealing with the possibility of a world behind our own whose events are decisive for the fate of our world as seen in the films *Stargate, Dark City,* and *The Truman Show*. It also reminds us of the allegory of the cave from Plato's *The Republic* (cf. chapter 9). Behind the artificial reality of the matrix lies a reality where forces are fighting to determine what is real, good, and true; they even determine the survival of humanity. Apparently, it is part of the worldview in *The Matrix* that an individual's course of life and subsequent fate are connected with a fundamental order of things, and that engagement with that world defines what the *human condition* could gain or lose by way of authenticity.

---

2   F. Rose, "The Second Coming of Philip K. Dick," *Wired* 11.12 (December 2003); see also www.wired.com/wired/archive/11.12/philip.html (retrieved March 19, 2011). The expression might have been taken from the postscript to Umberto Eco's *The Name of the Rose*. Eco explicitly connects this idea to postmodernism.

3   See also David Chalmers, "The Matrix as Metaphysics," in Christopher Grau (ed.), *Philosophers Explore* The Matrix (Oxford: Oxford University Press, 2005) 132–176.

This brings us to our second point. Not only does the film address all kinds of themes that deserve philosophical reflection, but different suggestions and alternatives for those questions are also presented for our consideration. What is being human really all about? What do people need or have to be in order to live an authentically human life? *The Matrix* offers a number of possibilities, generally embodied by different characters. What is indispensable for a full human life? Faith and hope (Morpheus, the god of dreams)? Power and survival (the Merovingian, Agent Smith)? Comfort and pleasure (Mouse, Cypher)? Self-sacrificing love (Neo, Trinity)? Does it come down to making choices (the Oracle)? And so on.

It is now a matter of discovering what is indispensable for being human. What does a person have to do or go through? We must examine the existential structures or vital features of humanness—questions that the *Matrix* films answers:

1. *Does it show in your corporality that you have (done) something truly human?*

    Councilor Hamann of the Zion Council simply points out the human need for water and air.

    Some suggest that human existence acquires meaning through pain and suffering. "As a species, humans define their reality through misery and suffering," Agent Smith says. When Neo is bleeding, the Merovingian comments: "See, he's just a man." After having been punched in the matrix, Neo looks in surprise at his own blood when he is back in the Nebuchadnezzar and says: "I thought it wasn't real."

    Does good food and pleasure define authentic humanness? Different scenes, both inside and outside of the matrix, show people eating with varying degrees of relish. The crew onboard the Nebuchadnezzar eat bowls of goop containing everything the body needs. The Merovingian owns a luxuriously decorated restaurant just for fun. And Cypher betrays his savior in exchange for a life in the matrix where cigars and red wine smell good and steaks taste delicious, even though it is only what the matrix tells his brain cells.

    Others suggest the significance of struggle and fighting. "You don't truly know a man until you fight him," says an oriental aphorism uttered by Seraph, the guardian of the Oracle. *The Matrix* contains a great many fight scenes, and Morpheus says of the ultimate battle: "This night holds the very meaning of our lives."

    Lastly, an individual's physical mode of existence is also indisputably expressed in dancing and sexuality. "To deny your own impulses is to deny the very thing that makes us human," Mouse says about the woman in the red dress he designed for the training program. Noteworthy are also the numerous nightclubs and discotheques in the matrix, and the grand celebration in Zion confirming Morpheus' speech that there is still human life.

2. *There is also the suggestion that the human condition is marked primarily by one's attitude to life.*

For some, this entails taking a leap of faith. "Morpheus was prepared to give his life because he believed in something," Neo says. Morpheus later says about Neo: "He's beginning to believe!" as a sign of the truth. This could even be seen as a recurring theme in the story.

Emotions and memories also play a role in the debate. Neo recalls eating good noodles somewhere, but this was an illusion. Yet Trinity says, "The matrix cannot tell you who you are." And later on, "The Oracle told me I would fall in love and that that man, the man that I loved would be the One." Her emotions are part of the truth.

It has also been argued that the ability to make choices defines human beings. "I don't like the thought of not being in control of my own life," Thomas Anderson says. According to Morpheus, "everything begins with choice." And, why does Neo persist until he has defeated Agent Smith? "Because I choose to."

Does being human live up to its potential by following one's vocation or destiny? "We're all here to do what we're all here to do," the Oracle says. The Keymaker shares the same life philosophy: "It is my purpose."

Can a person's *nobilitas* be found in hope and confidence? "Hope . . . the quintessential human delusion. It is at the same time the weakness and the strength of your species," says the Architect. But it brings Neo to free Morpheus, to rescue Trinity, and—finally—to defeat Agent Smith.

3. *A third group of arguments used in the anthropological debate originates from the world around us, providing the structures for human existence.*

People are embedded in a specific time and place. Morpheus points out that the surviving human beings have broken adrift in that respect: nobody knows exactly what year it is.

Another case in point is the moment in *The Matrix Revolutions* where Neo and Trinity soar high above the clouds and Trinity sees the real sun and moon for the very first time. Also, near the end, someone says of a breathtaking sunrise that it is so beautiful that it would have made Neo very happy. Clearly, these are anchoring points for existence.

Could having children be an essential anchoring point? Cas, one of the inhabitants of Zion, has two small children who give him joy and responsibility. The same is true for the mother of Jacob, a crewmember of the hovercraft Gnosis. The couple at the train station between the matrix and the machine world (in *The Matrix Revolutions*) also illustrates how precious their child Sati is to them. When Agent Smith destroys this child, the Oracle calls him "Bastard."

Reciprocating relationships of dependability and responsibility are essential in the films. The Oracle responds to Neo's suspicion that she might be a program in the system by saying that she is only interested in one thing: "the future. And the only way to get there is together." Similarly, Neo and Trinity can only make it together.

Lastly: Neo's negotiations with the machines also include an appeal to a higher order, an ancient force or "deep magic" (C.S. Lewis), forming the foundations of this world. This is a force that even the machines obey. An order where life is better than death, order better than chaos, peace better than war, faithfulness better than betrayal, persisting better than perishing, love better than hate, hope better than despair. Eventually even the Architect, who represents the "survival of the matrix," confirms this point. He vows to keep the promise that the machines will not destroy the humans, even if they want to be freed from the matrix.

In this first rough sketch of characters and events, we thus recognize a dialogue on the nature of humanity, our attitude towards this nature, and the different interpretations of both. It is therefore not surprising that the language employed in the *Matrix* films abounds with references to communities and traditions that embody and convey interpretative frameworks of humanity and world. Some of these will be addressed in subsequent chapters. By way of conclusion to this first exploration of *The Matrix*, we will briefly discuss these ideological elements.

3. The lion, the witch, and the matrix

Even the secular viewer should recognize that *The Matrix* trilogy is full of religious references. As has been said before, it should come as no surprise that religion is brought into play. The postmodern mindset implies that we no longer know which facts of life are still vital to human life, what approach to take to death, life, love, power, et cetera, and which interpretations of these existential facts could offer us firm soil. It is not surprising that a film that focuses on these issues also magnifies and puts to the test the meaning of traditional interpretations of our existence. This includes ideas from religions and philosophies such as Hinduism, Buddhism, Christianity, Gnosticism, and others.[4] Although *The Matrix* seems an eclectic rather than a Christian film,[5] we will confine ourselves here to a number of biblical references, peeling only the first layers off the surface.

In the first place, the religious references in the films are embodied by the characters. Neo is clearly a Christ figure; he is the new man. His name in the matrix is Thomas—known in the gospels as the doubter. The name Anderson literally means "son of man," a term used in the Bible for Jesus. Because of her inspirational role, Trinity reminds us of the Holy Spirit (the same can be said of the Oracle because of her advisory role) and as such completes the triad Neo–Morpheus–Trinity. On meeting her for the first time, Neo responds: "The one who has cracked the IRS d-base?" That is the one who knows all our *debts*. Morpheus is a kind of father figure (although not a creator, as opposed to the Architect), but also reminds us of John the Baptist, the Precursor. We also encounter a Satan figure, Agent Smith, and a Judas figure, Cypher. This is the first layer.

The characters are not the only ones who contain allusions to the Christian tradition; physical objects and locations also pertain to Christianity. Morpheus' ship, the Nebuchadnezzar, bears a biblical name as well, as does the last residence and destiny of free humans, Zion. The hovercraft's serial number, Mark III no. 11, refers to Mark 3:11, testifying that the Savior has come. A few more examples: in the first film, when Morpheus instructs Trinity to find Neo and free him, he sends them to "Adams Street Bridge"—the bridge in the way of Adam. The second film revolves around Neo's mission, assigned by the Oracle, to find the Source. This is the place where the matrix is created and maintained; in other words, this is where life originates. In the third film, when Trinity, Morpheus, and Seraph take the elevator to the restaurant of the Merovingian, they have to press the "help" button, but the lettering has worn away so it reads "hell."

Aside from names of persons and objects, we find a second layer of

---

4   See also the section on philosophy on www.whatisthematrix.com (also published as Christopher Grau (ed.), *Philosophers Explore* The Matrix (Oxford: Oxford University Press, 2005); and Saey and Garrett 2003; Michael Brannigan, "There is No Spoon: A Buddhist View," in William Irwin (ed.), *The Matrix and Philosophy* (Chicago: Open Court, 2002) 10ff.; and Gregory Bassham, 'The Religion of *The Matrix* and the Problem of Pluralism,' in Irwin 2002, 111ff.

5   See for instance Christian movie critic Steve Greydanus on *The Matrix* on his website: decentfilms.com/sections/reviews/matrix.html.

events that carries obvious Christian symbolism and metaphors. Smith explains somewhere to Morpheus that the first version of the matrix was perfect, but the humans refused to accept it, so a second, imperfect version followed. This seems to echo the biblical story of the Fall. One could even see Neo's liberation of Morpheus as alluding to the story of Jesus raising Lazarus from the grave. Once freed from the matrix, people are "in the world but not of the world," as the biblical expression goes. Lastly, Neo's sacrifice in the third film shows similarity with Christ's sacrifice for mankind; he is carried away in a crucified position.

In the following chapters we will approach some of these philosophical aspects of the *Matrix* films from a Christian perspective. It is now time to bring it to a close. If *The Matrix* can be called postmodern on account of its eclectic character, then it is probably more of a postmodern *Narnia* than a postmodern *Oz*. After all, the worldview of the main characters is based on the supposition that the world behind that of the matrix is "truly real" and that the powers and forces of the world behind our world are not only real but also bear relevance to our lives—a biblical picture. Choosing well is therefore our purpose, our responsibility (more on this in chapters 12 and 13). This is also the case in *Narnia*. The parallels were noticeable when the *Narnia* film was first released (Christmas 2005). In Frank Baum's story, the powerful Wizard of Oz does not actually exist; he is an impostor, and Oz as a land turns out to be a dream world. The main characters find their real selves purely by heading for the road; they achieve their destiny by their own doing, and are relieved when the world behind their own world appears to be an imaginary one. To interpret *The Matrix* in this way would not do credit to the film.

In this book we will make use of additional examples to demonstrate that it is no coincidence that both films refer to religious notions. It is debatable even with regard to cursing in the films whether or not it is simply a slip of the tongue—it only occurs in relation to Neo's character. The question remains, though, how exactly to interpret all these references. To illustrate this: those who watch *The Matrix* for the first time may experience the film as a great puzzle, and the biblical references may appear so profound that it seems undeniable we are dealing with the work of Christian filmmakers. However, when you watch it a second time and grasp the puzzle of the double reality, the biblical references may seem nothing more than embellishment. It might even lead to the conviction that the filmmakers hold the cynical view that religion only has a function as long as one doesn't understand the mystery. As Morpheus puts it, you can feel the matrix when you go to work, *when you go to church*, when you pay your taxes—it is all an illusion. But this is much less obvious again to those who have seen the film half a dozen times. Does illusion not only exist by the grace of reality? How else could we tell the difference? Perhaps the filmmakers offer the Christian interpretation of the human condition deliberately as one of the options, without passing judgment on it. This renders a response from us as Christian authors all the more challenging. We will go in search of the magical land behind *The Matrix*.

Chapter 6

TECHNOLOGY AS THREAT

> Neo: "What truth?"
> Morpheus: "That you are a slave, Neo.
> Like everyone else, you were born into bondage,
> kept inside a prison that you cannot smell, taste or touch.
> A prison for your mind."
> (*The Matrix*)

1. "The Matrix has you"

Neo has fallen asleep behind his computer. Suddenly the words on the screen are erased and a message appears, "Wake up, Neo!" Neo is roused from sleep and opens his eyes, clearly confused, and reads the words in astonishment. He gives the CTRL X command, but the letter "T" emerges. He tries another command, but this time an "H" appears. He randomly presses a number of keys, but without effect. The computer simply continues to type its message. Neo gazes at the four words on the screen, "The Matrix has you." He presses the ESC key and a new message pops up, "Follow the white rabbit." He hits the ESC key again. He rubs his eyes and sees that another message has appeared, "Knock, knock, Neo." At the same moment, someone knocks on his door. A young man named Choi and his friends have come to buy hacked information. After the deal has been closed, they invite Neo to go out with them. Neo notices the tattoo of a white rabbit on the shoulder of Choi's girlfriend and decides to join them. In the bar, Trinity addresses him and tells him that she knows why he can't sleep and why he spends his nights behind the computer. She knows that the question that has been haunting him has brought him to the nightclub, to her. He whispers, "What is the matrix?"

What the viewers do not realize at this point, but what they will soon find out is that *The Matrix* presents a society where humanity is entirely dominated by technology—by "the matrix." Indeed, it is a society where humans are reduced to a source of bio-energy. Only by virtue of the machines' compassion can they still enjoy life as a virtual reality. The movie thereby clearly and compellingly communicates a message about the threat of technology.

This chapter focuses in greater detail on the dominant character of technology as represented in *The Matrix* and other science fiction films (see previous chapters). Subsequently, we seek to achieve a better understanding of the phenomenon by drawing on the ideas of the philosophers Jacques Ellul, Langdon Winner, Henk van Riessen, and Egbert Schuurman.

2. Technology is taking over

The matrix is a computer-generated dream world, a projection of the world as it existed at the end of the twentieth century, that is inserted into the human brain. This illusion is pumped into the minds of millions. The matrix is a virtual world where people are born, live, and die. It is in this virtual world that they go to work, eat their meals, relax, and sleep (together).

Not surprisingly, Neo believes he lives in the year 1999. But Morpheus bursts his bubble by telling him they are actually closer to the year 2199. The situation is grave: earth is ruled by intelligent machines that control the majority of mankind. Only a fraction of the world's population lives in freedom in an enclave called Zion, deep under the surface of the earth. Most of these people are unwittingly held captive in liquid-filled containers that are connected to the matrix—the virtual world where people believe they live. The bodies of these "slaves" supply the matrix with electric signals. The matrix converts this input into signals that are transmitted to the brain. Their experiences are therefore virtual experiences. *The Matrix* constructs a scenario of the consequences of technological development that can be summarized as *technology is taking over*.

The Matrix also illustrates the notion that technology is an invisible force controlling us unconsciously. In the words of Morpheus, we are locked up in a prison that we cannot smell, taste, or touch. It is a prison for the mind. Apparently, individuals can become so engrossed by technology that they simply fail to see that their thoughts, feelings, and experiences are determined by technology. And the notion of "a prison for your mind" is frightening.

Films provide an instrument of reflection on this phenomenon as they, literally, appeal to the imagination. In 1927, German filmmaker Fritz Lang wrote and directed *Metropolis*, a film where technology is used to suppress a large part of the world's population, the "laborers." Almost a decade later in 1936, Charlie Chaplin produced *Modern Times*, where he offers a satirical and critical view of the influence of technology on humanity. Charlie's character is a tramp who works on an assembly line in a factory, constantly repeating the same routine. Slowly but surely, the role of machine is imposed on him. Later on in the movie, we see Charlie being used as a guinea pig, testing a new piece of machinery that automatically feeds the laborers, so they do not have to interrupt their work. In the final part, Charlie is sucked into one of the largest machines and gets trapped between its cogwheels.

The belief that technology will take over existence also figures into modern science fiction films. *Terminator I* (1984), *Terminator II* (1991), and *Terminator III* (2003), for instance, address the battle of humanity against intelligent machines. Other films, such as *WALL-E* and *The Day After Tomorrow,* deal with threats to the environment caused by the use of technology. Examples of this genre that we will consider are *Waterworld* (1995) and *The Postman* (1997), both featuring Kevin Costner. In some respects, the two movies can be regarded as opposites. In *Waterworld*, technology has resulted in the flooding of the world: the earth—excepting a few islands—is covered with water from the melted Antarctic ice sheets. *The Postman* shows how technology causes the dehydration of much of the earth in a nuclear war. The movies show similarity not only in their display of technology's destructive effects on the natural world, but also in the emergence of a Messiah-like figure, played by Costner. Reluctantly, he rekindles the hope for an earthly paradise. The film *The Day after Tomorrow* (2004) deals with rapid climate change generated by global warming and disruption of the thermohaline circulation in the Atlantic currents. After tornadoes have left a trail of destruction along the west coast of the United States, a cold front quickly moves in over the northern hemisphere and within minutes, the world is covered in snow and ice. The movie concentrates on the chaos these events produce in the major cities.

3. "What is the matrix?"

The Matrix depicts a society with highly developed technology, a society where anything seems possible. The movie presents a wide range of technological breakthroughs or "blessings":

- reproductive technology has advanced to a stage where sexuality and motherhood are no longer necessary;
- humanity has developed machines that surpass human intelligence;
- man has invented machines with consciousness;

- information, knowledge and skills can be inserted directly into the human brain;
- humanity can create a virtual world that is indistinguishable from the "real" world.

The Matrix also shows the price paid for these developments: technology is no longer in the service of mankind; instead, mankind is dominated by technology.

What is the matrix? *The Illustrated Contemporary Dictionary* lists the following meanings for the entry "matrix": "(1) That which contains and gives shape or form to anything. (2) A mold in which anything is cast or shaped. (3) A papier-maché, plaster, or other impression of a form from which a plate for printing may be made. [Latin: womb, breeding animal]."[1] These three meanings accurately express the message of *The Matrix*: we are in danger of shaping man according to the model of technology as a result of scientific progress. Metaphorically speaking, technology is regarded as the womb of a new human race.

What is the matrix? Read Mercer Schuchard provides a concise answer to this question. In his view, it is "the technological society come to its full fruition. It's Charlie Chaplin's *Modern Times* and Fritz Lang's *Metropolis* for the 21st century in which we don't simply work for the machine (rather than the machine working for us), but we are created, given life, and used by the machine exclusively for the machine's purposes."[2] *The Matrix* portrays a society that has realized the ideal of technological control, a society that no longer questions the nature and position of technology, and a culture where intelligent machines constitute the next step in evolution.

Does *The Matrix* offer a realistic account of technology? Is it possible to develop conscious machines? Will computers be more intelligent than human beings? Can information be directly uploaded to the human brain? Technologist Raymond Kurzweil holds the view that modern technology will fulfill all these promises.[3] Based on the rapid developments in digital technology, he estimates that *The Matrix* will become reality within the next three or four decades. He believes that the bulk of our mental processes will be "non-biological," i.e., sustained by electronic processes,

---

1   Chicago: Ferguson Publishing Company, 1978, 445.
2   Read Mercer Schuchard, "What is the matrix?" in Glenn Yeffeth (ed.), *Taking the Red Pill: Science, philosophy and religion in* The Matrix (Dallas: Benbella Books, 2003) 20.
3   Ray Kurzweil, "The Human Machine Merger: Are we heading toward The Matrix?" in Yeffeth 2003, 220ff.

by the year 2050. It will then also be possible to *download* knowledge and skills via bioports. As a result, the human mind will become a hybrid of biological and electronic thinking: "We will effectively merge with our technology."[4]

Kurzweil is optimistic about the immense possibilities of technology and proposes to "go with the flow." Remarkably, however, this is not the message conveyed by *The Matrix*. At the end of the movie, Neo urges the viewer to come into action and not let the matrix become reality. He says: "I believe deep down, we both want this world to change. I believe that the Matrix can remain our cage or it can become our chrysalis. . . . That to be truly free, you cannot change your cage; you have to change yourself."[5] This message corresponds with Neo's development in *The Matrix*: if you want to change reality, you will first have to change yourself. Initially, he did not believe that he was "the One," but that conviction gradually grew in him—and only then was he able to defeat his enemy Mr. Smith (see Ch. 4). In other words, as long as you do not transform yourself, the matrix will remain a cage that you cannot escape. It becomes a cocoon, a chrysalis, keeping you incarcerated.

What is the matrix? Schuchard underscores the moral claim of *The Matrix*, saying that "it is preaching a sermon to you from the only pulpit left."[6] The phrase "the only pulpit left" points to the film as the only remaining medium capable of getting through to the young generation.

4. "Out of control"

When watching *The Matrix* and other science fiction films, we are faced with a number of questions. What, exactly, is technology? Is it possible that technology will come to dominate us? What do we have to do to stay in control?

Philosopher Jacques Ellul (1912–1994) studied the character of modern technology intensively. The central notion in his analysis is the autonomy of technology. In other words, technology has developed into a powerful, independent force that increasingly redefines what it means to be human.

In *The Technological Society* (1964), Ellul offers a sociological approach to technology.[7] He focuses on the effect of technology on human

---

4   Kurzweil 2003, 234.
5   This is the text from the original script (Larry and Andy Wachowski, 2001, 219). The movie uses a slightly different text.
6   Read Mercer Schuchard 2003, 24.
7   See Carl Mitcham, *Thinking through Technology* (Chicago: University of Chicago Press, 1994).

behavior and interpersonal relations. In his view, human beings always make decisions within the structures of social reality. While social reality was defined by capital in the nineteenth century, the twentieth century witnessed a shift in favor of technology. The latter has become the dominant force in our culture, determining western social life to an ever-growing extent. A good case in point is the way that modern communication tools have stylized social interaction.

Ellul makes a clear distinction between traditional and modern technology. The crucial difference is the position of technology in society. Traditionally, technology was used only in certain sectors, and technical means were limited. Innovation took place locally, and people could decide whether or not they wanted to use it. This completely changed with the advent of modern technology. Ellul defines modern technology not as the sum total of machines or technologies but as the "totality of methods rationally arrived at and having an absolute efficiency in every field of human activity."

This definition is significant in a number of ways. First, it involves a "totality of rational methods." Secondly, the aim is to achieve an "absolute efficiency." And finally, modern technology is applied in "every field of human activity." The provision of power supply demonstrates the distinction between traditional and modern technology. Windmills and watermills used to be employed to grind corn for farmers in the region— a classic example of a local power source. Now, things are dramatically different. Power plants are the most important power source, generating and distributing electricity through a network of transmission lines. While industries use the electricity to grind corn, consumers find numerous other purposes, using the same power. It is literally used in "every field of human activity." A modern power plant is a component of a large network, rather than a single entity like the windmill. As part of a country's electric infrastructure system, the plant has to meet certain efficiency requirements, especially now that the energy markets are deregulated.

Ellul introduces several key characteristics of modern technology, and three will be discussed below. The first is what he terms "automatism of technical choice." Contemporary society considers problems as manageable, and every solution should be a technical one. Non-technical solutions—adopting a more modest lifestyle for instance—no longer seem to be accepted. Secondly, modern technology reinforces itself. Technical progress generates new technology, and these innovations are rapidly implemented in all segments of society. Technology accumulates at an exponential rate. Thirdly, technique has become autonomous and evolves according to its own laws. As such, technology determines our culture rather than the other way round.

Ellul owes his reputation to the notion that technique has become autonomous. Technology, seen as the totality of rational methods, has liberated itself from the network of social, economic, and political conditions. It has even become independent of moral and spiritual considerations. This technological environment, circumscribed by technical norms and values, has taken the place of humanity's natural habitat. In short, technology has spun "out of control," just as the Merovingian says of the human condition.

In *The Technological System* (1980), Ellul takes his philosophy one step further. He asserts that technology is not merely a dominant force; it has developed into a *system*. The different technical elements—methods, machinery, technologies—are not isolated phenomena that are to be judged on their own merits; rather, they constitute a unified set of interconnected components. The computer allows human beings to form an integral part of this system, and as such eradicate humans as independent subjects. Within this technological system, individuals can only pose technical questions and imagine solutions that are founded on technical processes. The dominant ethic is that of technical efficacy and technical efficiency.

Ellul elaborates this view in *The Technological Bluff* (1990), where he challenges society's strained expectations of technology as the panacea for all our problems. He also points out that the negative effects of technology are greatly underestimated. He refers to this as "bluff."

Needless to say, Ellul's approach does not provide clear guidelines for individuals to stay in control of technology. As a theologian and as a Christian, he makes an appeal to us to take our responsibility, but how are we supposed to take that responsibility if technology evolves autonomously?

5. Artifacts have politics![8]

Ellul's point of view is particularly relevant in connection with the issue of technological control. He convincingly argues that "the matrix" might actually become reality, explaining why and how. After all, doesn't his analysis show precisely that technological thinking moves toward domination per definition? Isn't the inevitable conclusion that ethical or spiritual considerations can no longer influence the autonomous development of our technological society? If these questions are answered affirmatively, what exactly is the value of Neo's message in the final scene?

---

8   A variation on the title of Winner's article "Do artifacts have politics?"

Political theorist Langdon Winner has also performed extensive research on the matter of the autonomy of technology.⁹ In *Autonomous Technology* (1977), he dismisses Ellul's technological determinism, arguing that Ellul does not do justice to the fact that human decisions always come into play in the evolution of technology. He focuses on the political dimension of technology in particular.

Winner wants to make clear that the autonomy ascribed to technology is not a quality of technology itself but of the product of human behavior. He uses Mary Shelley's *Frankenstein* to illustrate his view. What does Victor Frankenstein do when his creation has come to life? He runs away! He could have made a different choice: he could have adjusted his monster and tried to educate it. Instead, he flees. As Martijntje Smits writes, "The horrible deeds committed by the monster in the story can be traced back to Frankenstein's escape, as he fails to take responsibility for his creation."[10] This example serves to demonstrate how technology can develop into a monster if people do not concern themselves with the effects of technology on the environment and on society.

The question is how technological progress may be approached with necessary caution. Winner does not provide a clear answer in *Autonomous Technology*. He points out that the political nature of technology—denoting its capacity to impose a certain balance of power—is greatly underestimated. In subsequent publications, he emphasizes that political concerns should be explicitly addressed in engineering design processes. He believes that the power of technology can only be restrained if subjected to democratic control.

Winner thus made his most significant contribution to the debate on the autonomy of technology by acknowledging the political nature of technology. This acknowledgment renders it possible to overcome the opposition between technology and society. The notion that technology can be controlled in the political debate is based on the separation of the technological and the democratic domain. This, however, is a problematic notion, as Winner himself has shown that the two domains are inseparable.

6. "If you can free your mind"

Is it possible to overcome the opposition between technology and society? Could we enable a responsible development of technology while

---

9   See Mitcham (1994).

10 Martijntje Smits, "Langdon Winner: Technology as shadow constitution" in *American Philosophy of Technology: The empirical turn*. H. Achterhuis (ed.) Trans. Robert P. Crease. (Indiana University Press, Indiana 2001) 147–170.

acknowledging its political nature? In the final scenes of *The Matrix*, Neo answers these questions in the affirmative. He says: "But now, I see another world: a different world where all things are possible. A world of hope. Of peace." But how can we reach that new world? Neo's answer is hardly concrete: "I can't tell you how to get there, but I know if you can free your mind, you'll find the way."[11] The pith of the argument is described in the phrase "if you can free your mind." So, what does it mean?

We set out to answer these questions by drawing on the work of Reformational philosophers Henk van Riessen and Egbert Schuurman. In his book *Mondigheid en de machten* (Maturity and the Powers), Van Riessen discusses in great length the powers that threaten the maturity of modernity, such as freedom and responsibility. He refers to science, technology, and management as three notable powers that have been detached from and set off against humankind. As such, they threaten human maturity. Van Riessen identifies a dialectical tension: science, technology, and management have given individuals tremendous power over their environment, but these same fabrications have now turned against him and begun to dominate him. According to Van Riessen, this situation is the consequence of the "lapse of maturity," that is, a maturity that has wrested itself from God and his commandments. Technology and technical innovation are no longer in the service of God's Kingdom; they are used solely for the benefit of humanity.

Schuurman, a student of Van Riessen, expanded this analysis in his writings. In his dissertation,[12] he provides a philosophical study of modern technology and offers a liberating perspective. He takes up the theme again in the more recent *Faith and Hope in Technology*.[13] His criticism centers on the idea that technological development in our society takes place under an "empty heaven." In other words, western thought rejects the idea of living in a world created by God. Schuurman points out that an individual's pursuit of controlling reality through technology stems from a fundamental attitude that is best characterized by the phrase "lord and master." Individuals want to be master over reality and refuse to lay down authority. They seek to govern nature in order to create a future, to live in absolute freedom. In short, this still current attitude embodies the

---

11 Larry and Andy Wachowski 2001, 219.

12 His *Techniek en toekomst* (1972) was translated Donald Morton in the 1980s and recently reprinted as *Technology and the Future: A philosophical challenge* (Grand Rapids, MI: Paideia, 2009).

13 Schuurman, Egbert. *Faith and Hope in Technology* (Toronto: Clements, 2003) 67. Originally published as *Geloven in wetenschap en techniek. Hoop voor de toekomst* (Amsterdam: Buijten & Schipperheijn, 1998).

ideal of free and autonomous individuality.

Is this ideal of free and autonomous beings attainable? The Enlightenment—an eighteenth century philosophical movement in European culture that expressed a firm faith in reason—provided a clear answer to that question: the ideal of freedom and autonomy can be achieved through science and technology. In this manner, western individuals can safeguard themselves from the forces of nature and model society in a way that allows them to fashion their future without restraint (more on this utopian ideal in the following chapter).

The crux of the matter, however, is that the urge to control reality backfires. Science and technology have protected humankind from nature, but this protection comes at a high price: pollution and climate change demonstrate that humanity's natural environment is corrupted to the core. Science and technology have enabled individuals to cultivate themselves, but the cost is enormous: technology is increasingly taking hold of society, thereby severely restricting human freedom. Consider, for instance, the vast network of information systems—such as Facebook or LinkedIn—that allow personal data to be shared. As a result, a person's work, consumption, and recreational behavior can be mapped and manipulated by means of specific marketing methods. This is what Herman Dooyeweerd calls the conflict between nature and freedom. It is the paradox that humans are controlled by the very instruments they developed to control nature. Ironically, modern humanity has seemingly become a slave to the means that were meant to endorse freedom.

Schuurman uses the term "technicism" to describe the fundamental attitude of "enlightenment": "Technicism is the pretension of humans, as self-declared lords and masters using the scientific-technical method of control, to bend all of reality to their will in order to solve all problems, old and new, and to guarantee increasing material prosperity and progress. By means of their technology, humans want to control and safeguard the future. This technicism answers to two important norms as though they are the two great commandments: the norm of technical perfection or effectiveness and the economic norm of efficiency . . . . *Everything outside that narrow framework is denied recognition.*"[14] In other words, we will only find the way if we free our minds from the enlightened technicist state of mind.

7. "I see another world"

The Wachowski brothers never intended to make a neutral film.

---

14 Schuurman 2003, 69.

Their aim was to challenge people to think about life's great mysteries, including the meaning of life.[15] They hope to offer a new mythology based on a combination of the Jewish Christian faith, Gnosticism, Buddhism, and Greek philosophy. This explains why different motifs from these religions exist alongside each other. The Jewish Christian idea of the savior is embodied in Neo, who will save the faithful citizens of Zion from the power of machines.[16] But Neo's final words also point to the Buddhist notions of self-enlightenment and self-salvation. The amalgamation of these motifs creates a tension that increases when the Gnostic and Greek elements are taken into account, since the alleged radical nature of salvation is at odds with self-enlightenment. In other words, the idea that the use of technology leads to the subjection of an enlightened individual, leaves little hope of breaking the dominance in the path to self-enlightenment. Doesn't *The Matrix* show that mankind has come to a dead end? Doesn't *The Matrix* invite us to explore other ways?

A key concept in the philosophy of Van Riessen and Schuurman is the "being-as-meaning" character of reality, denoting that everything is created by God, exists in dependence of God, and is directed towards God. It entails that scientific and technological development ought to be aimed at serving God. This notion situates the evolution of science and technology in a normative framework that is diametrically opposed to the idea of a person as master. It is sustained by the belief that God's laws, as revealed in the Bible and as discovered through our exploration of reality,[17] provide an important guideline for the development of science and technology.

In the final scene of *The Matrix*, Neo envisions another world: a world of hope and peace. But, he was not the first to see it. Almost two thousand years before, the apostle John was given a vision of another world: a world without tears and sorrow, a world without sickness and death, a world without war and violence.[18] And this new world has a striking feature. A "voice from heaven" says to John: "Now the dwelling of God is with men, and he will live with them. They will be his people, and God himself will be with them and be their God."[19] Then, using the metaphor mentioned in the previous section, heaven will be empty indeed—as God lives with men on earth!

15   Saey and Garrett, 2003, 11.
16   We want to stress that this motif is not elaborated in a biblical way. The idea of "salvation through reconciliation" is entirely absent.
17   See the *Belgic Confession*, Article 2.
18   Revelation 21 and 22.
19   Revelation 21:3 NIV.

Chapter 7

TECHNOLOGY AS UTOPIA

> Cypher: "I know this steak doesn't exist.
> I know that when I put it in my mouth,
> the matrix is telling my brain that it is juicy and delicious."
> (*The Matrix*)
>
> Agent Smith: "Have you ever stood and marveled at its beauty?"
> (*The Matrix*)

1. "Not where, but when": No place

There is only one character in the *Matrix* trilogy who is really pleased with the technology featured in the films, and that is Cypher. Thanks to the technology of the matrix simulation he has the chance to enjoy a good steak, a welcomed change to the unappetizing food consumed day after day by the inhabitants of the "real" world. As far as he is concerned, the illusion of the technologically produced taste is preferable to the tasteless reality. Long live technology! Cypher's view obviously differs from the gloomy picture of the future that is painted by the trilogy as a whole. To Cypher, a life full of technology is not a dystopia (cf. previous chapter), but a utopia. Considering the literal meaning of the word, the matrix is a kind of utopia as well. "Utopia" is derived from the Greek "ou topos," meaning "no place." The matrix is no place; it is a computer simulation. And, in Cypher's opinion, it is an ideal world.

The transition from "ou topos" to "u-topia" was made by the sixteenth-century English philosopher Thomas More. In 1516, he published a novel entitled *Utopia*, which describes an ideal world. Aware that this world did not (yet) exist, he provisionally named it Utopia, a non-place. Poverty and misery were entirely banished from Utopia, and although it did not actually exist, as an ideal it was well worth exploring.

More was not the first to envision an ideal world. St. Augustine's *City of God* and Plato's *Republic* also contained elaborate descriptions of an ideal world. The idea of striving for utopia received great impetus during the Age of Enlightenment, a time marked by increasing optimism about the capacity of human reason. It was thought that Utopia might come within reach if humans would concentrate sufficiently on the power of their mental faculties. This optimism has its roots in the Renaissance. While Augustine's utopia is still dependent on God as its ultimate foundation, the Renaissance changed this attitude, taking humanity as focal point of a utopia. The humanist philosophical movement represented this attitude, and perhaps the humanist ideal has been of greater importance for the development of science fiction films than the fear of a dystopia.

In this chapter we will see how, according to science fiction films, technology might contribute to realizing a utopia. In the *Star Trek* television series and films in particular, it has been elaborated in different ways, and inspired by humanistic optimism as well. It is one of the few science fiction films that display an unequivocally positive view towards technology. (Steven Spielberg's *Close Encounters of the Third Kind* also comes to mind; but other than that . . . ?) This is reason enough to examine *Star Trek* in greater detail. Yet even here weapons can be found that one would not expect in a utopia. Responsible development of technology will take evil impulses into account. We will turn to Reformational philosophy for suggestions of such responsible development.

2. The endless frontier

Unbeknownst to many in the younger generations, there are several *Star Trek* series. It all started with what is now known as *The Original Series*. Its successor was called *The Next Generation*, which has since spawned a number of films. These were followed by *Deep Space Nine*, *Voyager*, and finally *Enterprise*, which is considered to be the last series. All episodes of the first three seasons have been released again on DVD. Watching them makes us realize that the film technology of the time seems almost endearing, as we are used to the impressive images of recent films like *Star Wars* or *Lord of the Rings*. Except for the fact that you don't see strings, there seems to be little difference with *Thunderbirds* (after all, *Star Trek* involves human actors rather than puppets). Still, it was quite an achievement back then. However, the series was not a success initially, and it was mainly owing to the later *Star Trek* films featuring the same actors that the phenomenon became what it is today. The word "fan club" does not begin to express what Trekkies are. These people are prepared to appear at conventions, to dress in *Star Trek* outfits, to wait hours to see their beloved heroes in the flesh.

Everything about *Star Trek* is politically correct. The captain of the

Enterprise, the star ship of *The Original Series*, is a white American named James (Jim) T. Kirk. The captain in the second series, Jean-Luc Picard is white and American, too, but he also clearly represents a particular minority group in America, namely, those of French origin. The next captain, Benjamin Sisko, is in command of space station Deep Space Nine. He is African American. The captain of the Voyager is a woman, Kathryn Janeway. It would have been perfect if the next captain were disabled, but in the *Enterprise* series a white American figures again as captain, whose name, Jonathan Archer, contains no reference to any minority group. This completes the circle, and the filmmakers apparently thought it would do: as far as film production company Paramount is concerned, the *Enterprise* will not get a successor.

Star Trek creator Gene Roddenberry always frankly admitted taking his own humanistic convictions as a starting point for the series.[1] *Star Trek* shows the beautiful things that can happen when a person is given space, both literally and figuratively. Roddenberry was personally involved in the production of *The Original Series*, but he died during the filming of *The Next Generation*. His humanistic ideas thus feature most prominently in the first two series, and although the makers of the later series claim to have carried on "in the spirit of Gene," a number of elements gradually crept in that Roddenberry himself would have excluded. In *The Original Series*, for example, every new and unfamiliar phenomenon is in the end carefully turned into a rationally controllable entity. But characters in different episodes of *Deep Space Nine* and *Voyager* even have religious or pseudo-religious experiences that are never rationally explained. Still, optimism about our capacity to build an ideal world, or at least to be heading in the right direction, is unmistakably a constant factor throughout the *Star Trek* series. Time and again, the crewmembers get to spread the good news among the extraterrestrials that hunger, poverty, disease, and war have been conquered on earth.

Roddenberry sought different ways to illustrate this in the composition and behavior of the *Star Trek* crew. Undoubtedly, Roddenberry had the diversity of races in mind when he created the wide variety of extraterrestrial beings. The crew of the famous starship Enterprise in *The Original Series* (and in the 2009 film *Star Trek*) demonstrates this. Beside Captain Kirk, it includes a black female communications officer (Lieutenant Uhura), a Chinese commander (Lieutenant Sulu), and an Eastern European communications officer (Pavel Chekov) was added to the cast

---

[1] Barad, Judith and Ed Robertson. *The Ethics of Star Trek* (New York: HarperCollins, 2000) xiv.

after complaints from the Soviets about the us/them mentality in the series (this was during the time of the Cold War). These human races are represented in all subsequent series. In the Voyager series an officer of Native American origin was added to the cast (Chakotay). The crewmembers have an exceptionally close relationship, and it grows stronger with each adventure they endure. It always takes extraterrestrial influence to incite crewmembers to quarrel.

The so-called "Federation of Planets" in the *Star Trek* series is an alliance between all life forms on the different planets. The aim of the Federation is to secure peace between the multitude of life forms. Owing to human good will, life forms begin rapidly joining the Federation. In the first series, the Klingons, an extraterrestrial warrior race, are still the Federation's greatest enemy, but in the second series (*The Next Generation*) one of the officers onboard captain Picard's starship is in fact a Klingon (Lieutenant Worf). *Andromeda*, another series that was written by Roddenberry and only after his death rewritten for television by his wife Majel, also focuses on the unification of life forms as humanity's highest purpose. The main character, Captain Dylan Hunt, shows such readiness to put all his own interests aside in order to achieve peace that he almost becomes a caricature. The humanistic ideal is magnified to such an extent that it becomes incredible. This may have caused the lack of popularity of the *Andromeda* series in comparison to *Star Trek*. The *Star Trek* series reveals something of the struggle it takes to pursue humanistic ideals. Yet even here we are presented with an extremely optimistic picture of the human capacity for good.

3. "Go where no man has gone before"

In *Star Trek*'s utopia, technology plays an important part. In the first place, technology enables contact between different life forms. Thanks to technology, humans can now "go where no man has gone before." The entire existence of starships revolves around exploring new life forms in order to enter into a peace bond with them. The technology of starships makes this possible. The ability to fly faster than the speed of light ("warp drive") is important. As soon as humans reach this stage in their technological development, they are approached by the Vulcans. This life form has long been able to travel faster than light and only makes contact with life forms that have developed the same ability. As made explicit in the Vulcan philosophy, this feat is considered a sign of civilization. The technological capability of a life form (mankind in this case) is taken as an indication of its level of sophistication. According to the Vulcans, savages will never achieve high technology. In this sense, technology is ethically charged: a high level of technology signifies that a person is highly civilized.

With regard to this technology, *Star Trek* constantly expresses high expectations. One example of this is the belief that at one point in time (i.e., in the time of *The Next Generation* and after), mankind will be capable of developing a kind of technology that possesses emotional faculties; humans could befriend this technology. One of the main characters in *The Next Generation* is Data, a humanoid robot whose character develops over the course of the episodes. Early on, the focus rests on his inability to have feelings and to understand the feelings of humans. But his desire to have feelings of his own intensifies as the seasons progress. In the feature film *Star Trek Generations* he appears to have an emotion chip that is normally switched off but is now activated at his request. As Data's functioning is severely disrupted by this change, it becomes clear that emotions are not a matter of an on/off switch, and it is decided it is better to de-activate the emotion chip.[2] Still, the longing for real human emotions does indicate that technology has got to a point where machines with emotions are within reach.

There are several other science fiction films that toy with this idea. In the "classic" science fiction film *2001: A Space Odyssey* (1968) the on-board computer HAL appears to experience feelings of fear when astronaut Steve decides to shut him down by removing his IC's one by one. The protagonist in Steven Spielberg's film *Artificial Intelligence* is an artificial "boy," struggling with his desire to be loved by his human mother. His desire is fulfilled in a moving final scene when his mother (or rather a clone of the mother, retrieved from the DNA extracted from a hair centuries later) explicitly declares her love for the boy. This declaration more than compensates for the fact that the clone only lives one day, as the AI boy has already reached his life's goal. The film *Bicentennial Man* (1999), based on a story by Isaac Asimov, also revolves around a robot trying to become as human as possible, even if it takes him 200 years.

In *Star Trek*, affection between AI characters and humans plays an important role as well. No matter how artificial Data may be, the rest of the crew develops a strong affection for him. His awkwardness in dealing with human emotions actually contributes to his charm. In the *Deep Space Nine* and *Voyager* series, we meet a hologram that even displays signs of intentionality: the Doctor. His grumpy disposition, manifested

---

2 Richard Hanley. *Is Data Human? The metaphysics of Star Trek* (New York: Basic Books, 1997) 112.

in his behavior from the outset (he complains every time people forget to turn him off after he has performed his medical procedures), may well be a matter of programming. However, in one of the first episodes he establishes with evident regret that he is the only crewmember without a name, thereby showing the first symptoms of emotion. One of the last episodes deals with the question whether he, as the author of a book, could legally be considered a "person" since it is a necessary status in order to claim copyright. Although it takes considerable effort for Captain Janeway to gather evidence, the verdict eventually says he possesses so many features of personality that he should indeed be treated as a legal person. The fact that Janeway spares no effort has everything to do with the strong affection the crew has developed for him despite his programmed grumpiness. All these examples demonstrate that science fiction films indeed hold such high expectations of technological developments that they even allow for the possibility of meaningful relationships with artificial beings complete with emotions.

4. **"It was a disaster": Cracks in the utopia?**

Weapon technology is unmistakably part of Roddenberry's vision of the future. And the weapons are used frequently, too. When their starship is seriously threatened, none of the *Star Trek* captains shrink back from completely destroying the hostile spacecraft. True, it is always for the purpose of self-defense rather than attack or revenge, but the outcome is the same. It seems that at the time *The Original Series* came into existence, Roddenberry agreed with the American government's policy of using all-destructive nuclear weapons as an instrument of deterrence. Apparently, he did not consider this incompatible with his humanistic ideals. Although humanism and pacifism often go hand in hand, for Roddenberry, the combination is by no means self-evident or imperative. In his view, weapons could also be an instrument in seeking peace. Individuals may protect their own lives, if necessary by killing opponents.

Seizing weapons only takes place in *Star Trek* after initial attempts at communication and negotiation have failed (invariably followed by "Beam me up, Scotty!"). Technology plays a vital role here, too. In space, just as on earth, different languages are spoken (yet we continually discover that the most peculiar creatures learn to speak perfect English soon after their first acquaintance with *Star Trek* crewmembers). Translation computers have reached a high level of linguistic competence in the *Star Trek* era. In the most recent series, *Enterprise*, we encounter the most prominent problems. *Enterprise* is chronologically set before *The Original*

Series, and all later techniques, including communication technology, are not quite developed yet. They still need a person with great linguistic skills (Hoshi Sato) to enable communication with other life forms. She does make use of a device that can generate an entire foreign language after registering a few sentences in that language, and which thereby functions as translation technology.

This technology is extremely valuable in *Star Trek*, as communication in itself is considered valuable for establishing peace and mutual understanding. The episode "Darmok" in *The Next Generation* is entirely devoted to this idea.[3] Captain Picard learns to communicate with a creature that works with words in a very associative way. Names from old sagas and legends are used as terms for contemporary concepts. Picard, therefore, has to study the cultural history of this life form in order to communicate, and this is exactly what Roddenberry regards as the ideal form of contact between cultures on earth. Humanism for him primarily means: being open to other ways of thinking.

The first most important law in the Federation ("the Prime Directive") states that humans shall never force their culture and norms upon other life forms. This seems a form of ethical relativism. Reality, however, is stubborn. Time and again the Prime Directive is trampled upon when the Star Trek crew encounters different cultural traditions that, by their standards, they see as unjust. This foregrounds an internal tension within humanism. Respecting all other opinions leads to internal conflict when confronted with someone else's claim to truth. It requires something of intolerance among the tolerant to prevent such a point of view from prevailing. Technology offers no way out of this paradox.

5. **Technology between utopia and dystopia**

It seems that individuals have always dreamed of an ideal world, and technology has come to play an increasingly explicit role in attaining that ideal world. Even in a well-elaborated humanistic ideal, as in *Star Trek*, technology does not bring these ideals any closer. Technology remains ambiguous even here: it can be used to connect people as well as to destroy them. Both in utopias and in dystopias it becomes clear what a powerful instrument technology can be in the hands of friend or foe. It is significant how little attention the *Star Trek* series gives to the extent that the nature of technology itself is dependent on norms. It is rare that the question is raised as to whether or not a particular technology should have been developed in the first place. Optimism about the possibilities

---

3   Barad and Robertson 2000, 29–35.

to use technology for good prevails entirely over the knowledge that that same technology could also be abused. The fact that technology opens up *both* possibilities is hardly ever mentioned.

The question of whether or not a specific technology should be developed is not an easy one, in particular from a Christian perspective. The question that comes into play here is how much power can safely be put into human hands. This arises as soon as knowledge—necessary for developing technology—is gained through research. Practice proves that it is almost impossible *not* to apply knowledge once it has been acquired. The world depicted in *Star Trek* is only too real in that respect. But there is also another side to the matter: we as humans are called to further cultivate our world and to gain the required knowledge by exploring reality. The crucial point is which norms and motives do we use in the process.

What does responsible development of technology encompass? Let us take a look at nanotechnology. It is not an unusual example in the context of a book on science fiction, as nanorobots ("nanobots") occur more and more in science fiction films and in television series. They never appear on screen; it is not even possible, because a "nanometer" equals one billionth of a meter, approximately the size of atoms. It is in fact a collective term rather than an actual technology. Almost all technology involving manipulation of individual atoms is referred to as nanotechnology. Besides, much of what is written about it deals with what might be possible in the distant future rather than what is currently within reach. Yet it is useful to not only consider the latter when making choices concerning nanotechnology. Our utopian dreams can also give rise to caution. In 1947, Thomas Watson, the founder of IBM, said that a few computers would suffice for the worldwide need for computing power. What experts call unrealistic today could well become a possibility in the near future. Utopias can be technically feasible, depending also on how hard we work to make the dream come true.

There are certainly dreams regarding nanotechnology. The ultimate "grail" is to build larger structures by snapping together individual atoms like Lego blocks. Eventually, nano gurus predict, we would be able to engineer human tissue and use it to replace dead tissue. We would also be able to build microscopic robots that can be inserted into the body in order to perform repairs independently. Similarly, however, we could create cameras of imperceptibly small dimensions and use them for all other kinds of purposes.

The main question therefore is: "what dreams come with this?" It could be the dream to find a more effective cure for certain diseases. It is a good thing if it stems from motives of love and care. But the dream of those who want to conflate humanity and machine (transhumanists) is to achieve immortality by replacing dying tissue just as often as we like (as in SF films *The Island* and *The Sixth Day*). As such, we would be able to reverse the ultimate punishment for sin: death. If this is really the motive, it means we would deliberately go against God's will. It means we would be building a new tower of Babel. Even in reducing suffering, we must be aware that we should not want to eliminate suffering at all costs. Francis Fukuyama, for example, has warned that humans could become vulnerable if they never experience any pain, and that technological development aimed at eliminating suffering will eventually be counterproductive.[4] The Bible also teaches us that suffering is used by God to accomplish something good (but it does not lead to the glorification of suffering). Several guru ideas on nanotechnology reveal the urge to completely control matter; control for the sake of control is not a good motive from a Christian point of view.

The first requirement from a Christian philosophical perspective is to use the right motives when making choices. Secondly, we must constantly ask ourselves how far we wish to go in developing methods to control nature. After all, we put all the knowledge and instruments of control in the hands of sinful humans. Working with imperceptibly small particles and structures offers unprecedented possibilities to do good, but also, e.g., to invade privacy. This information gives power. Undoubtedly, nanotechnology could provide terrorists with possibilities that make nuclear weapons look like child's play in comparison.

In the third place, responsible development of technology involves doing justice to the world God has given to us. Our reality is ordered according to certain laws that we have to respect.[5] A person is not a machine, and an animal is not a plant. Sometimes it seems nanotechnology does not differentiate between them because they can all be built from the same atoms. Much of what is written about nanotechnology therefore has a materialistic tendency. Human life and soul are reduced to material structures built with nanotechnology. This does not do justice to the differences God gives in his Word. Animals should not be treated as machines. Machines need *maintenance*, but animals need *care*. A utopia that is pursued without taking these distinctions into account turns into

---

4   Fukuyama, Francis. *Our Posthuman Future: Consequences of the biotechnology revolution* (New York: Farrar, Strauss and Giroux, 2002) 173 ("A person who has not confronted suffering or death has no depth").

5   Monsma, Stephen V. (ed.). *Responsible Technology* (Grand Rapids: Eerdmans, 1986) 65, in this context speaks of normative principles for development of technology.

a nightmare. Not everything that is currently beyond reach should be brought within reach.

6. "Live long and prosper": Towards a "eu-topia"

The full realization of a utopia is not likely to occur in our broken contemporary world, but the motives of love and servanthood are guidelines in that direction. We should also discern which forms of technology merit support. After all, we could, even with the best of intentions, set in motion a development that eventually leads to a dystopia. But without values like love, servanthood, justice, and peace, there are no good prospects for technology at all. In a Christian perspective, these values are not regarded as the fruit of human effort as is the case in a humanistic perspective. They come from beyond us, as a gift from heaven, as a source for the wisdom needed to determine which technologies these motives ought to lead. It is our human responsibility to seek that wisdom and to apply it to the development of technology, inevitably influencing our attitude to reality (see previous chapter). Such responsible use of technology offers prospects.

The end of the Bible does not present an "ou-topos" (a nonexistent place) but displays the reality of a new heaven and a new earth—a "eu-topia" (a good place). The new earth is described by means of a thoroughly technical image, namely the new Jerusalem. True, it is an image, but it is nonetheless significant that it is a technical one. It suggests that there is a future for technology on the new earth. So, there is a place for technology after all in what is *as yet* a "ou-topos," a place that does not yet exist.

Chapter 8

TECHNOLOGY BETWEEN REALITY AND PERCEPTION

Morpheus: "Welcome to the desert of the real."
(*The Matrix*)

1. "Electric signals interpreted by your brain": Real or hyper-real?
As shown in the previous chapters, the *Matrix* films contain all sorts of allusions to philosophical issues. But there is only one explicit reference to a philosopher in the entire trilogy. It occurs in an early scene in the first film when Neo is visited by other hackers and takes the disk containing the information they need from a hollow book. The camera briefly shows the title of the book, *Simulations and Simulacra,* by the French philosopher Jean Baudrillard, originally published in 1981 as *Simulacres et simulations.* Baudrillard shows in his book that our understanding of reality is more and more determined by artificially constructed images. He described the present-day United States as a desert where people only see reality to the extent that it appears in artificial images, as a kind of *fata morgana* (or mirage). Morpheus's famous statement "welcome to the desert of the real" possibly refers to that.

Baudrillard speaks in this context of hyper-reality. His criticism was initially directed at the way American television represents reality, showing that he was not merely referencing the techniques of "virtual reality" as seen in the *Matrix* films.[1] Think, for example, of the video images of missiles striking targets in Iraq, documenting the accuracy of American precision bombing. You cannot see what is *actually* happening; they might not have been strategic targets at all. Baudrillard not only discusses how technology shapes a virtual reality, but he also deals with the way

---

[1] It is debatable whether Andrew Gordon is right when he argues that Baudrillard confuses or fails to distinguish television and "virtual reality" techniques. Andrew Gordon, "The Matrix: Paradigm of postmodernism or intellectual poseur? (Part I)," in Glenn Yeffeth (Ed.), *Taking the Red Pill: Science, philosophy and religion in* The Matrix (Dallas: Benbella Books, 2003) 89.

technology shapes an image of the non-virtual reality. Due to the image-shaping function of technology, we are offered a certain view of reality that starts to lead a life of its own and thereby becomes a reality in itself. One is led to believe that the reality conveyed through technology is the actual reality. Much of what we take to be "nature" is, in fact, a stylized image of nature. The boundary between reality and hyper-reality is then erased. Baudrillard's ideas show affinity with postmodernism, which argues that we can construct reality to our own liking.[2] This is indeed possible, thanks to the images produced by technology. Whether you applaud or deplore this, the fact remains that it is possible and that it does happen.

In this chapter, we will look at what different philosophers of technology have said about image shaping through technology. The most elaborate view on the topic has been articulated by the American philosopher Don Ihde, who takes a positive stance on the role of technology. A more critical view is held by Hubert Dreyfus, who has written mainly about the Internet. In conclusion, we will see that the positive and the negative views propounded by the respective philosophers—both of whom were influenced by Martin Heidegger—require a broadening of perspectives, and Reformational philosophy offers that.

2. Jacked in: The body extended

Baudrillard was not the first to discuss the role of technology in relation to the way we perceive reality. Reflection on the role of technology also takes a prominent place in philosophical movements radically different from postmodernism: in phenomenology and in existentialism. No less a person than the famous German philosopher Martin Heidegger expressed his concern over the fact that our view of reality is increasingly shaped by technology. Heidegger argued that technology is the reason that we no longer see the intrinsic value of things, but instead consider reality as mere "raw material" that needs to be developed to gain real value. Technology is so all-pervasive that it becomes difficult to think differently. Heidegger was extremely pessimistic about whether there is still a way back. He stated in a now famous interview that "only a god can save us." Since he did not believe in the existence of such a god, the statement

---

[2] According to Dino Felluga, *The Matrix* propagates postmodern thought. Dino Felluga, "The Matrix: Paradigm of postmodernism or intellectual poseur? (Part II)," in Yeffeth 2003, 71. This is debatable however, because *The Matrix* does distinguish between reality and hyper-reality. David Weberman also has doubts about this. David Weberman, "*The Matrix* Simulation and the Postmodern Age," in William Irwin (ed.), The Matrix *and Philosophy* (Chicago: Open Court, 2002) 227.

actually means that he saw no way out. As is often the case in existentialism, he can hardly provide ethical guidelines for a good life within that technological world. Being aware of the problem is already a good start.

Heidegger was soon followed by others who reflected on the ways technology influences our views and understanding of reality. Don Ihde is probably the best-known philosopher of technology in this line of thought. His main interest is the way that technology determines our perception of reality. Phenomenologists all agree that you can never perceive the things themselves, but only the way they appear to us. And that is precisely the point where technology can exercise its influence. According to Ihde, technology can be inserted between the things and ourselves so as to alter the way things appear to us. Ihde distinguishes four different modes:[3]

- embodiment relations,
- hermeneutic relations,
- alterity relations, and
- background relations

We will further explain these in this chapter.

3. Through the looking glass

The first mode is described as an *embodiment relation*. Glasses are an example of this. Spectacle wearers do not see things directly, but through the lens of their glasses. Therefore, things appear differently to them than if they were not wearing glasses. In this case, it is most often considered an improvement, as they would see less clearly without the glasses. Still, we have grown so accustomed to looking through glasses that we almost fail to notice their presence. The glasses have become part of our eyes (which is what the term "embodiment" refers to). It is debatable whether glasses do, in fact, restore normal vision, as the image is distorted around the rim of the lenses. Conscious awareness of the glasses returns only when we notice these distortions by looking at something over the rim.

Ihde was probably inspired by one of Heidegger's examples. When a hammer is on the table, we can perceive it and be fully aware of its presence. But, if we pick up the hammer and use it to drive a nail into a piece of wood, the hammer seems to disappear from our consciousness and becomes one with the hand. Heidegger refers to this as a "Gestaltswitch." The hammer only returns to my awareness if I hit my thumb or if the head comes off. Only then do I become fully aware that I am

[3] Don Ihde, *Technology and the Lifeworld* (Bloomington: Indiana University Press, 1990).

holding a hammer.[4]

This phenomenon is also featured in *The Matrix*. The inhabitants of Zion are operating various mechanisms with long arms to fight off hostile machines. The movements of the mechanical arms are so natural and smooth that it seems as if they have become one with their operators. Only when an arm breaks down do they realize they were not actually fighting with their own arms. The hammer does not play an intermediary role in our perception of reality and, in that sense, cannot be compared to eye glasses. This also goes for Merleau-Ponty's example of a blind person for whom the walking stick almost becomes part of his arm and hand. But Ihde signals the same principle: it disappears from our conscious awareness. Glasses play an intermediary role in our perception, as does the hammer when it hits the nail.

Other visual aids also alter our perception, but they do more than restore normal vision. The microscope and telescope also function as intermediary between our eyes and the world that surrounds us, but they show things in a special way. As a result, we become aware—especially in the beginning—that we are not looking at the specimen or the stellar system with our naked eye but through the device. But even here the effect of the disappearing medium may present itself. When we peer through a microscope or telescope for a longer period of time and lose our thoughts in what we see, we forget that what we see we only see by virtue of the technical device.

There are some technical intermediaries that seemingly produce less distortion; for example, when one looks through a window to the world outside. But here too there are consequences: I do not see the whole of reality, but only the part that is visible through the window. In fact, I see a "framed" reality. The same thing happens when one looks at something through the eye of a camera. In former days this was a tiny window and it was very clear that one could see only part of one's surroundings: the window was used to "bracket" that part of the surroundings that one wanted in the picture. But this window has disappeared from most digital cameras, and most now have an LCD screen. According to Ihde, this relation is therefore different.

4. "See those birds?": Know what you see

The second type of intermediary relations between me and reality is what Ihde calls *hermeneutic* relations. It means that technology merges not with my sense organs but with the reality I am observing. The LCD

---

4   Ihde 1990, 31–33.

screen is a good example. Similarly, a power plant operator observes the power plant through its small digital indicators and display monitors, becoming one with the power plant.

As we can see in the *Matrix* films, the things that take place in the matrix can be followed on computer screens. Here, too, you are actually looking at a combination of the environment (in this case, a VR environment) and the representation of it as provided by a piece of technology. At a certain point, Cypher tells Neo that he has been working with matrix codes for so long that he no longer sees them as such. Instead, he sees what they mean right away: blondes, brunettes, redheads. Due to his extensive experience, Cypher immediately interprets these codes. Two other crew members, Tank and Link, see streets, cars, and people as soon as the green signs appear on the screen.

Another possibility is that the image produced by the device is so convincing that I forget that I am not looking directly at my surroundings but at a (technical) reproduction of it. A computer screen actually displays light and dark dots. The LCD screen of a digital camera provides a fairly accurate reproduction of reality. But that is not always the case, as other forms of photography illustrate. Infrared photography involves a modification of colors. It does not reproduce the colors I would see with the naked eye, but it indicates differences in temperature. An infrared photo of an avenue taken from the air therefore enables me to distinguish the trees that are dead, as these are colder than live ones. Color modification is also applied to pictures of stellar systems and planets. Similarly, the colors displayed are often not the ones we would see with the naked eye (for as far as we could, considering the distance), but they represent something else (e.g., in infrared photography).

When you are not aware that this modification has taken place, you could believe these are the actual colors of that particular heavenly body. You would forget that real, complete perception requires an interpretative move. You have to interpret the colors within a different context, for example in terms of temperature. That is why Ihde calls this the hermeneutic relation: technology renders an interpretation of reality, and we need to "translate it back." Another area where this often takes place is the medical world. Many modern technological devices that are used to make images for medical purposes require careful interpretation. Ihde mentions examples such as X-ray photographs, CAT scans, and MRIs.

5. **"Programs running all over the place": A new reality is formed**

A third type of the intermediary relations of technology is *alterity relations*, literally meaning that technology alters something in reality. In this sense it differs from the previous type, when the aim was to see reality *as it really is*, where the only role of technology is to help us see more or better than with the naked eye. But in this third type of relations, technology does not show us reality but something else. This happens when we watch a movie such as *The Matrix*. No matter how lifelike it may seem, we never see something that is actually happening at that moment. Reality may be implicit in the background: a documentary shows something that happened once or that is happening somewhere else. A film may even display the reality of the filmmaker's thoughts, like the ideas of the Wachowski brothers about reality. But they always deal with a reality that lies outside our direct perception. This type of relations could be used, however, to reinforce or extend our image of reality, as is the case in hermeneutic relations.

Although it is not science fiction, the film *Travelling Birds* is an impressive example of alterity relations. The filmmakers developed a miniature helicopter that they send to fly with a flock of migrating birds. They worked with endless patience to get the birds accustomed to the presence of the device so it could fly along with them without scaring them away. As a result of this process, the viewer seems to be flying with the birds. You can see their wings flap up close, and you instantly feel as if you are one of them. You can see the beautiful landscapes below as if through their eyes, occasionally looking sideways to the birds around you. It seems a different reality (hence, alterity relations), namely the birds' reality, all thanks to the technical intermediaries of a camera and a mini-helicopter.

6. "I wrote it myself": Beautiful images...

The fourth type of relations Ihde distinguishes is called *background relations*, meaning that the perception of reality is influenced by a technical alteration in the background. Ihde uses an example that involves hearing rather than seeing: the hum of the refrigerator. This sound is constantly audible in the background of all the other kitchen sounds. As a result, what we hear is different from what we would hear if the refrigerator was not there. Central heating is another case in point, providing a constant temperature (rather than sound) in the background. Similarly, there are certain smells that are produced by things around us that we (now) fail to even notice. Over the course of time we no longer focus our attention on these sounds and smells, and the phenomenon fades into the background (hence the name). In this case, technology does not alter the information sent from reality to our sense organs but adds something that in itself is not interesting enough to be registered (unless it gets too loud and starts to annoy us).

In contrast to his teacher Heidegger, Ihde's view of the intermediary role of technology is not a negative one. Rather, he acknowledges the enhancement of perception that comes with this role. We get to see things we would not normally see, and that can truly enrich our view of reality. At the same time, we should remain conscious of the role of technology. If not, we might create a wrong view of reality for ourselves, as is the case when we adopt an incorrect interpretation in hermeneutic relations. This will lead to a deceptive difference between illusion and reality (see next chapter).

Technology offers a wide array of intermediations between us and reality. This holds for each of the three types of relations described above. Alterity relations have developed quickly in the last decades, mainly due to the computer. Image editing, as made possible by computer software, enables us to produce images that seem surprisingly (and deceptively) real. Digital photographs can be edited in every possible way and, as a result, reality is represented in a distorted way. Through photographic image enhancers, we can "add" people to a picture who were not even in the scene when the image was taken. The consequences are many, including the fact that photographs can no longer be used as evidence in a criminal trial, unless it has been meticulously examined in order to rule out photo editing.

There are several other science fiction films beside *The Matrix* that deal with this notion, showing the negative sides of these technological possibilities. An interesting case in point is the film *Final Cut*, where people can have a memory chip implanted in their body that will record everything one perceives (visually and aurally at least). The implant can be removed after a person's death, and the footage can be edited into a film, to be used, for example, at the funeral to keep the memory alive. "Cutters," the people in charge of editing all the images stored on the implant, produce such films on their computers. *Final Cut* makes clear that footage can be manipulated in such a way that memories that trigger negative thoughts about that person can be excluded from the cutter's selection. As such, you are able to create an image of that person that can distort reality to a considerable extent.

The film *Simone* exemplifies another aspect of image editing techniques. The main character is a film director who received a disk from a deceased friend containing a program that allows him to insert a virtual actress, Simone, into any film. This is actually a double alterity relation: the film itself is a modified reality, and another modification takes place because Simone, unlike the other actors in the film, is a simulation: Sim-1. This film also emphasizes the negative rather than the positive effects of technology. Simone becomes insanely popular, and everyone wants to see her in person. For a while, the director manages to keep up appearances by creating a holograph of Simone suggesting she is actually present at the press conference or at the stadium where she is supposedly performing. Eventually, however, he is exposed as a fraud.

A final example is *The Truman Show*, where the protagonist leads his everyday life until he discovers by accident that he is actually the main character in a television show where his entire life is watched by an audience. He does not live in the normal world but in a gigantic studio, and he only discovers the edge at the end of the film. It concludes by Truman stepping through a door in the set, and straight into the real world. The constructed reality is not a virtual one in the sense of a computer animation, but it comes close. Although he is physically real, Truman's life on the show is a consciously constructed illusion, and for a long time, he is unaware of the artificiality of his world. As soon as he realizes what is going on, he refuses to put up with it any longer. He does not rest until he has escaped from the set. His plans of escape show that there is nothing like full reality, being a *true man*. It is possible that this movie inspired the makers of "reality television" programs such as Big Brother. But in those programs the players are aware of the fact that they are watched by an audience. In the "Truman Show" this is not the case.

7. "The matrix cannot tell you what you are": . . . not so beautiful after all?

The notion of the intermediary role of technology in our perception of reality is also exactly the message of another philosopher in the line of Heidegger, Hubert Dreyfus. He argues that this role of technology in our perception of reality is often an impoverishment. The Internet is a preeminent example. On the Internet, we can "visit" countries we would otherwise never have experienced. We can communicate with people with whom we would otherwise never have communicated. But in this case, the contact between us and reality takes place entirely on the mental level. There is no physical contact whatsoever. We cannot feel the grass or breathe in the air of the country we look up on the Internet. We cannot shake hands with someone we e-mail. In that sense, we are dealing with a strongly reduced form of contact with reality. Dreyfus states that this is insufficient for human beings, who not only have a mental, but also a physical side.[5] Neglect of the physical aspect of communication leads to an impoverishment of our experience of reality. In case of "virtual reality" there may be stopgaps like gloves with motion sensors and actuators, but these are still poor solutions. Dreyfus also applies this to education. Of course it is a good thing if you can use distance education to reach people who would otherwise have been destitute of education (for example in third world countries), but, Dreyfus says, we must never forget there is nothing like the physical presence of a teacher. That is just the way people are built, and we should do justice to that.

Reformational philosophy readily agrees.[6] Herman Dooyeweerd explicitly rejected the dualistic conviction that human beings can be divided into a physical and a mental part. The experience of reality involves our entire being; and when that is not the case, it is a reduced experience. Such a reduction is wrong, however, only when it turns into a reductionism. At that point we forget that we have "flattened" matters and take something for full reality when it is not. This happens, for example, when the results of one or another science, which always focuses only on one or two aspects of the whole, generate insightful but nonetheless limited knowledge that is so easily assumed to be a description of the whole of reality. Also in our pre-scientific experience of the world around us, we so easily presume to know true reality while having only experienced a certain side of it. Honed insight gains weight only when it contributes to a better understanding of the full experience of reality. This is also holds for Facebook, Hyves, and other social media. One can have a thousand friends on Facebook and yet be desperately lonely.

Phenomenologists like Ihde and Dreyfus also tend to embrace a reduction, namely, one's personal perception and experience of reality. Whether or not technology is good is then measured primarily by the question of how technology affects that perception or experience. In terms of Dooyeweerd's aspects, this entails first and foremost the examination of the psychological aspect (cf. chapter 3). But, one could also look at the social aspect. One's attitude towards the Internet may be more positive when taking into consideration that this medium offers opportunities for communication that were not there before, albeit in a limited way as Dreyfus correctly observes. Of course it is nice to meet

---

5   Hubert L. Dreyfus, *On the Internet*. London/New York: Routledge, 2001, in particular chapter 3.

6   Egbert Schuurman's critique of the Internet in *Faith and Hope in Technology* (Toronto: Clements Publishing, 2003) is similar to that of Dreyfus. He notes that the Internet should be linked back to the fullness of reality.

someone face to face, but that is simply impossible when the one lives in the United States and the other in the Netherlands. Another aspect to focus on is the economic. Resources are saved when a group of people use teleconferencing instead of traveling from all corners of the world to one place for a meeting.

An assessment of the Internet that is broader than one that merely emphasizes the psychical aspect could soften Dreyfus's negative judgment considerably. Conversely, a broader view can also lead to a qualification of the rather positive view taken by Ihde. From a legal perspective, one might not be inclined to welcome the opportunities technological devices offer to convert reality into a new reality. In such cases, how can we still distinguish illusion from reality? In short, if we want to reach a balanced judgment of the way that technology shapes images, we need a better view of the multifaceted norm-relatedness of technology than what phenomenologists have to offer.

Chapter 9

ILLUSION AND REALITY

Morpheus: "How do you define real?"
(*The Matrix*)

Cypher: "Real is just another four-letter word."
(*The Matrix*)

1. **"I thought it wasn't real": How real is "real"?**
Several science fiction movies play with the idea that we might be in a virtual world rather than in the physical world where we think we are. *The Matrix* trilogy, *The Thirteenth Floor, Vanilla Sky, Total Recall,* and several other movies are based on this theme and for each of the characters who suddenly find out that things are different than they had always believed, it is quite a shock to realize the new status of their existence.

Apparently, reaching "enlightenment" is no picnic. It means going through the depths, exactly as the process is depicted in various religions. Reaching enlightenment implies going through a crisis, but it is worth the trouble. A considerable adjustment was required before Neo could comprehend what real life is. When he was on the operating table, just after awakening in the real world, Morpheus told him that he had never really used his muscles or his eyes before. His whole life he had been nothing more than a motionless body in an artificial womb (as explained before in relation to the literal meaning of the word "matrix"). The real world appears to have become a dark place where thinking machines are in control and where humans are exploited as mere energy source. Only Neo's brain was active, and his body was merely sending energy to an installation providing machines with electricity. After awakening in his "bathtub," Neo saw other people motionlessly connected to the same installation, completely unaware of their actual state of being just as he was himself before Morpheus offered him the red pill. What he took for real life all along is now nothing more than a computer simulation—the matrix—where the machines conveyed messages to his brain. Indeed, this is no small shock to him or to us as viewers.

The question whether we really exist as we think we do (or is it all an illusion?) has occupied the minds of philosophers for centuries. In this chapter, we will go back in time and see how old this question really is. Beginning with a contemporary philosopher, Hilary Putnam, we will work back chronologically via Rene Descartes in order to arrive at good old Plato. The different viewpoints have been labeled in various ways over the course of history. The conviction that there is a world that is independent of our thoughts is often referred to as *realism*, and the belief that there is (perhaps) not such a knowable world is called *skepticism* (see chapter 5).

We need to be careful here, as a conflation of two different questions often takes place. First, there is the matter of *being*, the ontological question as to whether the world that I think I see around me is really there. Secondly, there is the matter of our *knowledge* of reality, the epistemological question as to whether I can obtain reliable knowledge about reality, supposing that it exists.[1] For that reason, we speak of ontic realism and epistemic realism, respectively. Both questions can be answered with "yes," but both questions have also been answered with "no." The two questions are closely connected: "if there is nothing, there is nothing to know, and if there is something, how much can we know?" It is difficult, however, to demonstrate unequivocally why and how both questions may be answered affirmatively. *The Matrix* makes this very clear: the fact that you experience your everyday life as real, as Neo did before his "enlightenment," is absolutely no guarantee that it is, indeed, real.

There are yet other questions that are connected to these two questions; for example: "Is there an objective truth that is the same for everyone?" And eventually there is a fundamental question like: "Can we ever know for certain that God exists?" It is in vogue these days to say that truth is a matter of perspective and experience. In *postmodern* thought, as this is generally called, everyone has his or her own truth. Postmodernists also apply this notion to the second question. It is fine if you want to believe that God really exists, but you should not expect everyone to accept this as a universal truth. Perhaps a "god" experience is nothing more than a subroutine in the matrix simulation. Philosophers have always seen the question of the existence of reality and of the possibility of knowledge about that reality as a challenge. And it is not an easy one, as this chapter

---

[1] The second question is also discussed in chapter 8. The latter mainly focuses on the question of how technology influences our knowledge of reality, while this chapter tackles the question as to whether we can gain reliable knowledge about reality at all.

will show. Reformational philosophy has also tackled this question. At the end of this chapter, we will explain why a Christian perspective provides more anchorage than those of Putnam, Descartes, or Plato.

2. Coppertop: Are you more than a brain in a vat?

The *Matrix* filmmakers, in what really amounts to a thought experiment, were not the first to tackle the question of the realness of reality. At least one philosopher, using an image almost identical to theirs, preceded them. In his 1981 book *Reason, Truth, and History*, Hilary Putnam discusses a problem that has become known as the "brain in a vat" problem.[2]

Stephen Law explains the thought experiment in a straightforward manner in *The Philosophy Gym*, a popular accessible book on philosophical issues.[3] He presents the problem in the form of a story. Imagine waking up one morning and seeing a strange creature, an evil extraterrestrial scientist who tells you he is using you for an experiment. He has removed your brain from your body during the night and placed it in a vat with life-sustaining liquid. Your brain is now wired to a computer that provides it with all kinds of impulses, making you believe you are still in your own body. He has just connected you to a machine that replaces your eyes, allowing you to see him. He turns the machine around, and, to your great astonishment, you see a vat with a brain wired to a computer. Your first thought is: "This can't be true. I'm having a nightmare!" But the mad scientist smiles and asks you why that would be the case. What arguments do you really have to establish that the "brain in a vat" scenario is not real and that your everyday life *is* real life?

Law undermines the arguments one after another. He refutes the argument that it would be a strange coincidence if we would all have the same bad dream when discussing this problem with the counter-argument that all our discussion partners might just as well be part of our own dream. One argument might be that other people tell us they see the same world we do. But Law refutes this with the counter-argument that we have no guarantee whatsoever that the others are really there and that the discussion is really taking place. The chapter ends with an open ending: can you ever be sure that you are not in a "brain in a vat" situation?

The similarity with Neo's story is so striking that it is very possible

---

2   Putnam's "brain-in-a-vat" discussion is related to *The Matrix* by Gerald J. Erion and Barry Smith, "Skepticism, Morality, and *The Matrix*," in William Irwin (ed.), *The Matrix and Philosophy* (Chicago: Open Court, 2002) 21, and Glenn Yeffeth, "*The Matrix* Glossary," in Glenn Yeffeth (ed.), *Taking the Red Pill: Science, philosophy and religion in The Matrix* (Dallas: Benbella Books, 2003) 246.

3   Stephen Law, *The Philosophy Gym* (New York: Thomas Dunne, 2003) chapter 3.

that the filmmakers used the story as source of inspiration, although they never refer to Hilary Putnam. There is a difference, however, between the way that Putnam describes the "brain in a vat" problem and Law's popularization of it. Law leaves the solution open, while Putnam uses the problem to show that our everyday life is real.

In that respect, Law's story is not so much reflected in the *Matrix* trilogy as it is in another science fiction film called *eXistenZ* by David Cronenburg (1999). This film resembles the theme of the *Matrix* in many respects. It also deals with people whose bodies can be connected to a computer via a so-called bio-port and thereby enter a simulation program. The virtual reality game in *eXistenZ* allows several people to be plugged in simultaneously, and they need to work together to get from one level to the next. The movie is quite violent in that several murders are committed as part of the "games." Perhaps the film implicitly criticizes the proliferation of violent computer games available these days. Killing opponents is the rule rather than the exception in these games.

The film starts in a hall filled with people, several of whom are seated on a stage in order to demonstrate the new game. After they have been plugged into the game, we follow them on their way through the simulation, which does not always result in pretty pictures, due to the amount of violence. Every now and then, the scenery suddenly changes, bringing us to another level. Near the end of the film, we are back in the hall where it all started. The film seems to be coming to a close, but suddenly someone in the audience gets up and shoots one of the game participants. One of the other participants looks at us in bewilderment and asks: "Are we still in the game or not?" All of a sudden we realize that the participants might not have come back from the game, but have simply reached another level. The film ends here, leaving the viewers confused about the answer to that question. We have seen so many scene shifts that we simply do not know. That is exactly the purpose of the film: to leave us with the question whether "real" is truly real.

In contrast to Law (and *eXistenZ*), Putnam's argument about the "brain in a vat" results in the conclusion that reality cannot be like that. His line of reasoning is not easy, and it is beyond the scope of this book to fully explain it. The main thrust of the argument is that the ability to conceive of the "brain in a vat" scenario (without question, an ability we all have) requires that we are not a "brain in a vat" ourselves. Because, if we were, the statement "I am a brain in a vat" would only allow us to think of virtual brains and vats. Since real brains and vats do not exist for us in that scenario, we cannot have a conception of what a "brain in a vat" would look or feel like. But then the statement "I am just a brain in a vat" no longer has the same meaning as in the original story, which *does* refer to a real brain and a real vat. So, according to Putnam, there is a valid line of reasoning that leads to the conclusion that we are not "brains in vats."

It is important to note that Putnam's line of reasoning involves the question of whether or not we can sensibly state that we believe ourselves to be a brain in a vat. Putnam's argumentation serves to demonstrate that we can know that we are not. His discussion thus addresses an epistemic question rather than an ontic one. His line of reasoning is therefore primarily considered as an argument against skepticism, the conviction that we cannot know anything with certainty (this is true at least for some things). Note also that he strives to set up his argumentation entirely through the internal consistency of statements. There seems to be no presupposition whatsoever about being or knowing. Others, like Law, have tried to undermine his argument by showing that it contains hidden presuppositions. For it seems that, from the very beginning, Putnam assumes the existence of real brains and real vats (for he presumes that we should distinguish between real and non-real brains and vats). It appears that, without ontic presuppositions, it is difficult to prove that there is a reality outside of us about which we can acquire knowledge. The Christian philosopher Alvin Plantinga, for example, points out that it is not possible to make a claim about what counts as rational knowledge without making a claim about whether our cognitive faculties are functioning properly. And this question leads to an area of anthropological assumptions concerning the question as to when a human being as a whole is functioning properly.[4] Law, therefore, concludes with an open ending: we simply cannot prove anything without presuppositions.

Others, such as Descartes, will not settle for that.

3. "Your mind makes it real": Finding anchorage in reason?

The setting of the *Matrix* films differs from that of *eXistenZ*. In *The Matrix*, it is constantly made clear if we are dreaming or waking: in the Matrix simulation, the scenery is usually light, and Neo and his friends are wearing long leather coats; the real world is always wrapped in dusky light, and Neo and his friends are wearing clothes made of rough T-shirt-like fabric. This pattern is never broken. The film thus emphasizes that after reaching "enlightenment" you will always know exactly where you

---

4   Alvin Plantinga, *Warrant and Proper Function* (Oxford: Oxford University Press, 1993) 214–215.

are: in the real world or in the virtual world of the matrix. In that sense we do not even have to wait for the "happy ending" of the *Matrix* trilogy to feel better than at the end of *eXistenZ*, which leaves us in uncertainty. In the movie *Inception* similar "signals" are used to indicate the difference between reality and virtuality (the individual carries a top totem to remind him that he is not being controlled; only he knows of the totem so that he can remain in reality).

It is not a particularly pleasant thought to not know whether or not our everyday life is real. Apparently, people are looking for a definite answer as to what is real and what is not because what we want most is to live in a real world. The philosopher Robert Nozick expressed the problem as follows. Even if someone was plugged into an "experience machine" giving him all the pleasurable experiences he desires, he would still prefer reality to illusion, even if the latter induces a more pleasant experience.[5] We would rather have Holland really win the World Cup, than "feel what it is like" in Nozick's machine for only a moment. According to Nozick, we do not only want to experience things, but we actually want to perform and savor them; we do not consider a mass of brain floating in a tank to be a real "person." Moreover, we do not want to be restricted to a reality that does not reach beyond what we made of it. This influences the freedom we enjoy (see next chapter). It is important that we know that reality exists. But how can we find out?

Ages before Putnam, another philosopher tried to find grounds to overcome his doubts about his existence: René Descartes. Everyone knows the famous phrase: "I think, therefore I am." Descartes formulated this phrase as part of a philosophical exercise in order to find an indisputable criterion for real knowledge, from the depth of his doubts, when most everything was uncertain. He systematically challenged all his certainties: he questioned the existence of his surroundings, the existence of his body, the existence of God, even his own existence—who's to say that he had not been deceived about a number of things, perhaps by an evil demon that makes him believe lies. In short: he faced the same kind of questions as Neo did in the first *Matrix* film. But, Descartes realizes that he himself was having these doubts. How could he have doubts (and therefore: think) if he does not exist? In this manner he regains his first certainty: *I know this much: I exist.*

---

5   Nozick is referred to in connection with *The Matrix* by Erion and Smith in Irwin 2002, 24; Theodore Schick, "Fate, Freedom and Foreknowledge," in Irwin 2002, 89; Lyle Zynda, "Was Cypher Right (part II)? The nature of reality and why it matters," in Yeffeth 2003, 42; and Peter J. Boettke, "Human Freedom and the Red Pill," in Yeffeth 2003, 148.

From that point, he climbs back up. Because, if he himself exists, God must exist. Where else would he come from? And if God exists, then all other things must exist as well because God is the source of these things too, and God is a God of truth. In the same way, he regains certainty about his physical existence. But if Descartes were to see the *Matrix* trilogy, he would undoubtedly start doubting whether his reasoning was valid: someone unknowingly floating in a liquid-filled tub, connected to the matrix is not much of a thinking "ego."

In any case, Descartes bases his belief in the existence of things on reason. He is, therefore, considered one of the founders of the Enlightenment. By the standards employed in the Enlightenment, only what is in congruence with the extent of our rational capacities can be accepted as knowledge. In Descartes' view, the existence of God still falls into this category, but that is no longer the case for a later Enlightenment philosopher, Immanuel Kant. He acknowledges that God exists, but he claims that belief in God is faith not knowledge. Real knowledge only comes from a combination of perception and the use of concepts that are already in our minds. According to Kant, knowledge of God can never meet that requirement: we cannot perceive him directly, and he is not a category of the understanding either. Hence, it is not real knowledge. If he is right, Descartes' elegant argument, intended to provide new certainty about the realness of reality, collapses at an early stage.

This is the second example of a failed attempt to reason about reality without making presuppositions. Someone once said that the whole of western philosophy consists of a series of footnotes to Plato. Perhaps it is useful to go back to the old master himself.

4. "A prison for your mind": The allegory of the matrix

Like Putnam and Descartes, Plato is often referred to in essays on the *Matrix* films, mainly in relation to the allegory of the cave that is found in *The Republic* (book VII).[6] Plato describes there a group of people sitting chained to the floor of a cave. They watch the moving shadows projected on the wall (owing to a fire they cannot see). The people have lived in the cave their entire lives and have no idea that those shadows are actually cast on the wall by real moving objects; they believe the shadows to be reality. Plato proceeds to describe how one of them escapes from the cave. His eyes have great difficulty getting used to the light, just like Neo struggled when first using his eyes. But once he reaches "enlightenment"

---

6   The most elaborate discussion is by Charles L. Griswold, "Happiness and Cypher's Choice: Is Ignorance Bliss?" in Irwin 2002, 128–129.

he realizes that what he took for reality all along is nothing more than shadow play. He knows better now and, like Morpheus, Plato believes that in that position he has the responsibility to return to his former fellow sufferers to tell them about the reality he himself has experienced.

It seems that Plato can be considered a realist. But we forget that his tale is intended as an allegory. What, then, does Plato's parable mean? Surprisingly enough, it is the opposite of realism. Because, for Plato, the "enlightenment" of the philosopher consists in his increasing awareness that the reality we see around us is only a poor copy of the ideal reality. This ideal reality, which we can deduce from perceptible reality through philosophizing, is the really real reality. In that reality, all circles are perfect. However, we cannot know that reality through sense perception but only through contemplation. Even when you are doing your utmost to draw a perfect circle, on close inspection you will always see that it is not perfect. You will therefore never discover all the properties of a circle by studying your hand drawn circle. Only mathematical reflection can teach you what these properties are. And that is true for all the other things you see around you. In our reality, motion is never perfectly steady, as there will always be a bit of shaking and shuddering. Only by contemplating the ideal reality can we discover what motion at a constant speed is. The ideal reality, not the perceived reality, is the actual reality.

The release from the cave thus brings Plato to something very different from the release from the "cave" of the matrix. His position on the reality of knowledge leads to a position on the reality of being: what we experience around us is not as real as what we can "perceive" through thinking. Should we abandon our desire to know the truth about the reality that we experience in everyday life because that reality is not worthwhile? This question has been answered affirmatively within Christian circles in the past: "It is not about this world." The immaterial and eternal soul rather than the material body is what really matters. However, this notion is in conflict with the way the Bible speaks about the body and the material world, namely, as God's good creation. With good reason we are given the promise that we will become men and women of flesh and blood once again, after death and after the Second Coming. Certainly, the body will be endowed with qualities it did not have before—after all, it is called the glorified body—but it is indeed a body, not a ghostly apparition. So, Plato's idealism is not the solution either.

5. *Reformed revolutions*

In the quest for certainty about the existence of an external world and the quest for reliable knowledge about that reality, Reformational philosophy goes its own way. Its uniqueness rests mainly in the connection between the question of being and the question of knowledge. For Putnam and Descartes, the answer to the second question determines the answer to the former question: we can only know if there is a reality outside of us if we can gain knowledge about it. They tried to acquire this knowledge through clever thinking, and we have seen they only partially succeed because they tried to do so without presuppositions. Kant does subscribe to the view that reality exists insofar as we know it, but he holds that we cannot know this reality without making *a priori* presuppositions. What we perceive has no real meaning if we do not use a set of concepts already present in our minds (for example, the relationship between cause and effect).

Herman Dooyeweerd turned the issue upside down by demonstrating that the problem has not been formulated correctly. No wonder, then, that the quest produces no satisfactory results. According to Dooyeweerd, we will never reach certainty about reality if we keep reasoning from our knowledge of that reality. He claims that the guarantee for the existence of reality cannot be found in our knowledge but in the fact that God created that reality. This is, of course, a principle of faith. But one may wonder if the view of Putnam, Descartes, and Kant is not a principle of faith as well. After all, they put their faith in human reason as the foundation for certainty about reality. That is no less an assumption than the Christian principle of faith.

The second flaw in the argument, according to Dooyeweerd, rests in the implied opposition between myself as knowing subject and reality as the object of my knowledge. The definition of the problem suggests that there is no direct connection between the two, but a gap that I need to bridge through reasoning. Dooyeweerd argues, however, that the chasm between subject and object does not even exist because I and everything around me belongs to the same created reality (see chapter 3). And in our everyday "naïve" experience of reality, we gain knowledge of reality directly. Only when one starts to analyze or philosophize about that reality does something of an "opposition" emerge. That "standing over-against" becomes problematic only if you forget that you can always return to the direct experience of reality.

In short, once you have overcome the initial shock of the uncertainty about reality brought about by the *Matrix* films, you can breathe a sigh of relief when realizing that things are not really that way. God placed us in a reality that we can observe and know directly, a reality that does not need further questioning. It can be a useful philosophical exercise to explore how our knowledge is connected to that reality, but there is no reason to doubt that we can know anything about that reality. The fact that we cannot figure it out by using reason alone does not mean that the uncertainty remains. There is a fixed point where we can anchor our knowledge. Dooyeweerd referred to this as the "archimedean" point (cf. chapter 3) in connection with Archimedes' statement that he could move the world with a lever if only he had a fixed place to stand. That one fixed point anchors our entire thinking and forms a key concept in Dooyeweerd's philosophy. Without it, we will never escape from our "Cartesian" doubt. We must seek that point outside of our own thought. God himself guarantees the existence of a richly varied and yet orderly reality. Both the variation and the orderliness stand open to our knowledge. God thereby personally provides a guarantee that there is a reality and that we can acquire knowledge about it. It is extremely fascinating to explore that reality through our direct experience as well as through more distant, abstract scientific knowledge.

Chapter 10

FREEDOM, DETERMINISM, OR DESTINY?

The Oracle: "Would you still have broken the vase if I hadn't said anything?"
(*The Matrix*)

The Merovingian: "You see, there is only one constant, one universal;
it is the only real truth—causality.
Action, reaction; cause and effect."
(*The Matrix Reloaded*)

1. **"The problem is choice"**
One of the recurrent themes throughout the *Matrix* trilogy is the question of whether we humans are free to make our own choices or if all our actions are determined by causes outside our control. The films present different attitudes towards this matter. Perhaps the Merovingian is the most explicit in expressing his beliefs. According to him, all actions are a matter of cause and effect. "Causality . . . we are forever slaves to it," is how the Merovingian summarizes life in *The Matrix*. He also draws the consequences: "Our task is not to create the future, but our only hope, our only peace is to understand it." Understanding is only possible when we gain insight into the causal relationships that determine the course of things. (And if the Merovingian is consistent in his theory, he will realize that understanding the future is also predetermined.)

Although Morpheus is less extreme in his comments on the subject, he also refers to an inevitable destiny on several occasions. He believes that Neo is destined to save the human race from the machines, and for that reason, he will certainly do so. Still, the existential struggle is most prominently felt by Neo himself. Is he really destined to save humanity as a messiah? Does he have the choice to do it or not to do it? These questions creep up on him especially when he is talking to the Oracle. She seems to know exactly what he is going to do. Does her foreknowledge imply that these things will have to take place, independent of Neo's own choice?

With good reason, the filmmakers connect the problem of foreknowledge with that of freedom of choice (see next chapter). There is also a theological dimension to this philosophical question. Is it possible to say that God is all-knowing in the sense that he knows in advance everything that is going to happen, and that we still have to choose for ourselves whether we want to follow him or not?[1] This philosophical exercise may become a vital question, just as it was for Neo. Of all the philosophical questions raised in the *Matrix* films, this is probably the one most liable to make you queasy. For most people, the question whether the world around us really exists or exists only in our minds is a challenging philosophical exercise rather than a reason to keep one awake at night. But how the question whether we are truly free to decide about our life's course relates to the question if that is all determined beforehand is not only conceptually difficult to resolve—it affects us in our heart as well. It is not for nothing that Neo grapples with this: is my role as savior predestined by the programmer of the matrix, or can I decide for myself? In fact, it is also the only question that the filmmakers leave entirely open.

This chapter discusses different views on freedom. We start out with the boldest position on the pre-determinedness of the course of things: "hard" determinism. Then we will take a look at the softer versions of determinism. Daniel Dennett is perhaps the most well-known philosopher who elaborated on this. Since this position does not yield satisfactory results from a Christian point-of-view, we will conclude by exploring the way Reformational philosophy tackles this problem.

2. "Only one constant": Hard determinism

In the *Matrix* films, the Merovingian is the representative of a philosophical movement called "hard determinism."[2] This position holds that everything, including human action, is a matter of cause/effect mechanisms. All events have an identifiable cause and can be fully explained by these causes. Note that only physical events are acknowledged as causes. In the most popular version of hard determinism, the laws of physics constitute the cause/effect mechanisms.

---

1 Theodore Schick, "Faith, Freedom, and Foreknowledge," in William Irwin (ed.) *The Matrix and Philosophy*, (Chicago: Open Court, 2002) 92, also makes the connection between the *Matrix*-films and Christian theology.

2 Matt Lawrence, *Like a Splinter in your Mind: The philosophy behind* The Matrix (Oxford: Blackwell Publishing, 2004) 59–62; cf. Mark Rowlands, *The Philosopher at the End of the Universe* (London: Ebury Press, 2003) 121ff.

The reduction of complex phenomena to simple physical laws can be a step-by-step process. War breaking out (a sociological phenomenon) can be traced back to the mental condition of individuals (a psychological phenomenon), which is a matter of the functioning of their brain (a biological phenomenon), which, in turn, is a matter of chemical reactions (a chemical phenomenon), which, ultimately, is a matter of atomic movements. And, vice versa: if we could determine the position and speed of every elementary particle at a given moment, we would be able to calculate what the world looks like for every following moment, thanks to the laws of physics.

It does not make much of a difference to refute this by saying that quantum physics taught us that we can never simultaneously determine the position and speed of any particle. This merely indicates that we may be able to predict the future not deductively but statistically. It is a difference in degree rather than in kind: the notion of pre-determination is not eradicated.

To support this view, one must assume that material realty is all there is; for that is what the laws of physics deal with. Any immaterial reality acting independently of physical substances would undermine a full cause/effect description. We will return to this point later in this chapter. Still, materialism has always had its followers, and determinism has also managed to survive the course of time.

There is a significant objection to hard determinism, however, as it does not offer a satisfactory answer to the question why we are still under the strong impression that we can indeed make free choices. It is incontestable that, while you are reading this book, you cannot help thinking that you are free to decide whether to continue reading or to stop at the end of this sentence and close the book. So where does that idea come from, if everything is predetermined? In order to do justice to the idea without immediately dismissing determinism, present day philosopher Daniel Dennett developed an interesting idea: it depends on the level you want to look at it.[3]

3. "As long as they were given a choice": Freedom as perspective

At first sight, Dennett's view seems to be in line with determinism. He believes that, on the level of physical phenomena, everything is a matter of cause and effect: this is, where the laws of physics are at work. The same is true for the level of biological life, where everything is determined by evolution. The process of chance variations and selection of the best of those is no less a matter of cause and effect than a matter of the phys-

---

3   Daniel Dennett, *Freedom Evolves* (London: Penguin Book, 2004).

ics of dead matter. Dennett agrees on this point with Richard Dawkins when he writes that if there is such a "watchmaker" (someone who has developed the world of living beings as a kind of watchmaker), he must be blind (i.e., he has no consciousness, no intentions). The "blind watchmaker" is no more than a metaphor for the blind evolutionary process. If we could determine the position and the speed of all elementary particles at any given moment, the laws of physics would allow us to calculate what the world looks like for every following moment, as we have seen in the previous section.

But, Dennett says, this is an entirely theoretical statement since we will never reach that point. Besides, we must ask ourselves if it gets us anywhere. It is such a huge step from all those individual molecules to the macroscopic world of things and humans that we can hardly imagine the practical meaning of the results of our immense particle calculation. Even if we were capable of making such a complex deterministic calculation, it yields little insight into the things around us. Moreover, we are left with a persistent feeling that we do have a say in things.

Dennett argues that it is more practical to take a different stance on the macro situation. He refers to this as the intentional stance, as opposed to the physical stance that focuses on the individual particles. This descriptive level allows us to discuss consciousness and intentions in an appropriate and reasonable way. It is also at this level that we sense that possibility of deciding for ourselves. These issues are not in play at the physical level, where you find only particles void of intentions and emotions. According to Dennett, that physical level is the actual level of being: there is nothing besides those particles. The intentional stance only denotes a clever way to talk about things on a higher aggregation level; in other words, it is a different language game.

So, intention does not actually exist; therefore, neither does free will. Sometimes, however, it is useful to talk about things as if intention and free will do exist. Besides, the "particle reality" is of such complexity that it really seems intentionality must play a role. Dennett explains this by drawing on John Conway's computer program *Game of Life*. This program takes place on a grid-like board with live ("populated") and dead (empty) cells. There are a few simple rules to get from one situation to the next: a live cell without live neighbors dies (as if caused by loneliness), a live cell with more than four live neighbors dies (due to overcrowding), a dead cell with three live neighbors in the following situation becomes a live cell. Due to these simple rules, the game is completely determined. But it is far from dull. Over time, the game's fans have discovered all kinds of little "creatures" with remarkable "behavior." The "glider," for example, is a pattern that travels across the board. There are also "creatures" that "devour" other creatures on their way. Another example is the "gun" that keeps "shooting bullets."

We use the double quotes to indicate that these board figures are not really creatures. It only seems that way because their form returns in each new situation, albeit in a slightly different place. Dennett would say that, physically speaking, we are dealing merely with separate cells or pixels that blindly obey the deterministic rules. But, thanks to the complexity that the game allows, it appears to us as if there are creatures showing certain behavior. Intentionally speaking, you could talk about creatures that are moving and devouring other creatures. We do that to simplify the descriptions, but they are really nothing but pixels that turn black or white in accordance with the rules. It is also a nice example of Don Ihde's "hermeneutic relation" (see Ch. 8). The intentional can be said to evolve from the increasingly complex game board. The title of one of Dennett's famous books also refers to that evolution: *Freedom Evolves*. Freedom is a phenomenon that takes place on the intentional level; it evolves from the complexity of the changing things (that change only in line with cause/effect relations).

On the surface this sounds like a plausible description, and many people are attracted by it. It is a solution that ties in with materialism on the one hand, and also does justice to our idea of freedom of action on the other. Yet it also provokes resistance, since from this perspective free will is merely an illusion. We talk about it because it simplifies the description, but in reality free will does not actually exist. Could that really be true?

Dennett's response to this objection is simply: what difference does it make? Why should we worry about the fact that our sense of free action is just a matter of point of view? Doesn't that sense work in daily life without causing any problems? So what is the advantage, Dennett argues, of saying that freedom of action does not exist, if you could not have acted differently than you did? Of course you could argue that in theory you had the freedom to stop reading at the point where you were invited to choose to do so. But you did not, did you? You continued reading, and now you have arrived here. So what is the use in saying you really could have stopped? Intentionally speaking, you are not at all disturbed by the fact that free will does not exist. After all, that is the level where we live and act, so what is the point in worrying?

Dennett has no difficulty ignoring our intuition on this point. The question is if doing so does justice to the value of our intuition—for it is our intuition that rouses our resistance when reading Dennett's ideas. This resistance cannot be pushed away as easily as Dennett would lead us to believe. Neo struggles with the same question as well.

4. "Because I choose to": How about responsibility?

The question of responsibility versus (pre)determinism is illustrated in the science fiction film *Minority Report* (2002). The question is answered in a remarkable way: people are convicted for crimes they would have committed if they had not been prevented from doing so. Three young people who are held in a liquid suspension chamber in a comatose state experience visions of future murders. Thanks to the special psychic abilities of these "precogs," policemen can see that a certain person is going to commit a murder at a certain moment in the future. As a result, they can go to the location in question just before the crime is committed and arrest the criminal before he can actually carry out his plan. The mere planning of the crime is thus marked as a punishable crime. That differs from our current legislation. Although planning a crime is not guilt-free, it is not punished in the same way as the actual crime. Moreover, even the crime that will spontaneously arise in the moment (and therefore has not been planned with malice aforethought) is considered as a full crime in *Minority Report*. The policemen in the film do not catch criminals in the act but before it; yet they are punished for a crime that has not been committed.

But can someone be held responsible for a crime that he was destined to commit (considering the fact that others have seen the crime being committed in the future)? Main character John Anderton, a police officer who himself is accused of a future crime by the system, finds himself in a situation where he is supposed to commit the murder. But he manages to break away from his "fate" by not doing what was predicted. The filmmakers thus chose against complete determinism for the sake of saving responsibility. But it was a close call. Or maybe even worse, because the film does not resolve the ambiguity on this point. In the visions of the "precogs," you can see the main character shooting, and you can see the other person being hit, but you cannot see whether it is really the bullet from the main character's gun that kills the other person. So, in theory, it is possible that the prediction is correct; thus, that determinism is real. It is possible that the filmmakers deliberately left it open to discussion.

Dennett might counter the responsibility question by saying that jurisdiction does not take place on the physical level but on the intentional one. And on that level, you may safely speak of intentions, and therefore also of responsibility. That these are a matter of "stance" does not make any difference. But it is clear that the *Matrix* films and *Minority Report* both grapple with Dennett's simple solution. Both films show that we do not easily settle for that; the struggle is far too real.

Dennett's solution is not satisfactory from a Christian perspective either, which says that responsibility is not a language game we can talk ourselves into by reasoning from a certain point of view but something that comes from God. He made us as creatures that can answer him when he asks: "Adam, where are you?" Nowhere in the first chapters of Genesis are we given the impression that he did not really have a choice whether or not to eat from the tree of good and evil. He is held responsible for the consequences, and not just in the first chapters of Genesis but throughout the Bible.

It is true that the Bible speaks of God's hand in our choices, especially concerning the most decisive choice in life: for or against God. But the Bible gives a twofold answer here. On the one hand, we read about humanity's own responsibility. We are called upon to mend our ways: "Do this and you will live." "Then choose for yourselves this day whom you will serve." "Continue to work out your salvation with fear and trembling." On the other hand, especially in Paul's letters, we see that God is needed to bring people to choose for him. In the Bible, our incapacity to make that choice is attributed to the consequences of the Fall. By having once made the wrong choice, we voluntarily made ourselves slaves (to sin). God needs to save us from slavery, but somehow this does not happen outside of us. Reformed theology has always kept aloof from the idea that God and humanity are cooperating in this. But that does not mean that we do not play a role. More than once, we read about the possibility of people refusing God's saving hand. "God does not rape us," the reformed theologian Van Ruler once wrote. That, of course, is not a fine philosophical solution to the problem of determinism versus free will. Maybe we should be prepared for never finding such a solution.

5. "Don't worry about the vase": Where does foreknowledge come from?

So let's go along with Neo, for whom real freedom of choice is still impossible to give up (we will return to this in the last chapter). But that conviction comes at a price. Every time he meets the Oracle, he faces a problem: if he really has freedom of choice, how is it possible that the

Oracle knows in advance what he will do? A few dialogues from *The Matrix Reloaded* illustrate his dilemma.

> Oracle: Why don't you come and have a sit this time.
> Neo: Maybe I'll stand.
> Oracle: Suit yourself.
> (Neo sits down)
> Neo: I felt like sitting.
> Oracle: I know. . . . Candy?
> Neo: You already know if I'm going to take it?
> Oracle: I wouldn't be much of an Oracle if I didn't.
> Neo: But if you already know, how can I make a choice?

Neo is clearly ambivalent about the Oracle's ability to predict his decisions. On the one hand, he does not want to waste energy on escaping his destiny if that is indeed impossible. On the other hand, he resists the idea that he would be a will-less object of that destiny. That resistance is so substantial that he would probably not settle for Dennett's idea that it is not so bad to be a will-less object of destiny (or the Architect) as long as you feel like you can choose freely. This stance also leaves the unsatisfactory feeling that the notion of responsibility is almost emptied of its meaning. How can you be held responsible for doing things you were destined to do without having a say in it? Then again, how does the Oracle knows what he is going to do?

The same question is brought up in relation to God's providence and his unconditional election. Generations of philosophers and theologians have tackled the issue of how these two are related. Some of them solve the dilemma by focusing on God's directing hand in events in an almost deterministic way in order to do justice to God's omniscience. Others turn to human freedom by renouncing the idea that God fully knows and controls things. To many people, this option is particularly appealing with regard to the question of how evil can exist. Neither approach is fully convincing. Perhaps the *Matrix* trilogy is more biblical in this sense than many a theologian, since the tension is never really resolved.[4] To the very last minute, Neo is torn between the predictions of the Oracle and the—apparently real—possibilities to go his own way. It does become clear that the Architect did not program him completely. It seems as if the Architect himself is also surprised by Neo's course of action. Towards the end of *The Matrix Reloaded*, he has a conversation with

---

4   Idem according to Michael McKenna, "Neo's freedom… whoa!" in Christopher Grau (ed.), *Philosophers Explore The Matrix* (Oxford: Oxford University Press, 2005) 234.

Neo where he admits that he had to reprogram the matrix twice because things did not go the way he planned. The Oracle does not always seem quite sure, either. Why is this tension not resolved, and how is it possible that foreknowledge and destiny in some way go together? It is interesting that both programs have to make choices.

6. "Have you always known?": Limited knowledge

The question of knowledge plays a vital role in Reformational philosophy. Perhaps one of the most important points of interest is that of the limitations and normativity of our knowledge. It pertains to the matter of freedom versus determinism in a number of ways, and, related to that, the matter of foreknowledge versus destiny.

In the first place, we should distinguish between the different kinds of cause/effect relations. Reformational philosophy speaks about different aspects of reality, each with its own specific set of laws (see chapter 3). There are relations that can be described in terms of physical laws; for example, in physics, we study the laws of the physical aspect. When I let go of an object, it will fall down to the ground. There is no intention of any kind involved on the part of the falling object. But there are also psychic "laws" that are studied in psychology when examining the laws of the psychic aspect. Those laws for instance say that people with certain character traits are inclined to react in a certain way in a certain situation ("I *knew* you were going to say that"). But they can still refuse to surrender to their impulses. Such behavior is too inconsistent to be explained from the position of hard determinism.

The distinction between knowledge related to physical processes and the entirely different kind of knowledge of psychological processes, where the relationship between cause and effect is completely different, is not only visible in science; we also know it from experience. Sometimes you just "know" what the other person is going to do. Is it because you can predict it based on physical cause/effect relations? No, such determinism is not in play here. Yet such a prediction is often correct, because you know that other person in particular and the patterns of the psyche in general. Could it have gone differently? Probably, but there seem to be certain laws at work. And this is with good reason, because life would not be livable without such laws. Those laws make reality "trustworthy," even if we do not know all the causes.

Another notion in Reformational philosophy that might be of even greater importance to the awareness of our knowledge's limitations is that we are under not above the laws. All the regularities we find in reality (what we call "laws" and are studied in different scientific disciplines) have been placed there by God. He sustains the regularities, thereby showing us his faithfulness. Because of this, he himself is above these "laws." We, however, are under the laws, and we cannot look "over" them. We can gather knowledge about these "laws," and we take every opportunity to do so. But why God does things the way he does—also when he sometimes decides to depart from these "laws"—far exceeds our comprehension (unless he expressly reveals it to us). Because of that, because we do not know everything, it is possible that we have to make a choice.

Lastly, there is the fact that God is above time, and we are in time. We can only think in terms of "before" and "after." Even when we contemplate foreknowledge and predestination (election), our understanding goes no further than "before" and "after." But God stands above time. He is not bound by "before" and "after." Apparently, the result of that is that his foreknowledge does not take away our freedom of choice, or reduce it to "fake" freedom. At this point we should embrace Calvin's attitude; he said we would be overestimating our intellectual capacities if we tried to find a fitting explanation. It is more significant to speak of God's providence in the sense that he provides for us. Abraham experienced how God provided an animal to be sacrificed in Isaac's place. God's care for us is not primarily revealed in his ability to see into our future but in that he wants to provide what we need for that future. What the future holds and in what way he has a hand in it remains hidden to us. It seems that we will just have to live with that. The difference with Dennett is that our responsibility is not a mirage (because everything is determined), but real responsibility because God asks us to act according to his will with our limited knowledge.

Unsatisfactory? Philosophically, yes. The most you can do is ask if the question is the right one. The fact that a question cannot be adequately answered might be a sign that the question itself is flawed. Maybe something is wrong with the expression "freedom of choice." If it is supposed to indicate a freedom that exists outside God's influence, then what exactly does that imply? Is that not the same as wondering how a square circle can be both round and square-shaped at the same time? Doesn't the problem arise because the question introduces an incorrect notion? It might get you further philosophically, but the existential struggle remains. Dennett's solution does not convince, and the philosophical solution probably doesn't either due to the incorrectness of the notion of freedom that exists outside of God. In the Bible, as well as in our own experience, that tension between foreknowledge, determinism, and freedom of choice continues to exist. But, then again, that is what makes life interesting. The makers of the *Matrix* films cleverly exploited that idea by refusing to resolve the tension between foreknowledge, determinism, and freedom of choice. And with good reason.

Chapter 11

CAN PROGRAMS MAKE MORAL JUDGMENTS?

> Agent Smith: "Why Mr. Anderson? . . . Why get up? Why keep fighting?
> Do you think you're fighting *for* something—for more than survival?
> Can you tell me what it is? Do you even know?
> Is it freedom, or truth, perhaps peace, could it be love?
> Illusions Mr. Anderson, vagaries of perception.
> Temporary constructs of a feeble human intellect trying desperately
> to justify an existence that is without meaning or purpose.
> And all of them as artificial as The Matrix itself."
> (*The Matrix Revolutions*)

1. "Why Mr. Anderson?"

The final confrontation between Agent Smith and Neo takes place in the closing scenes of *The Matrix Revolutions*, the third and last film of the trilogy. They try to defeat each other in a breathtaking fight on the streets but also in the air. It seems that neither party manages to gain the upper hand. When Agent Smith asks Neo why he keeps fighting, he expresses the view that, with the exception of the fight for existence, there is no reason at all to fight. In his view, values such as freedom, peace, and love are illusions of a weak human mind.

Why does Agent Smith say that (moral) values are an illusion? Is it because he is programmed that way? We could put these questions in a more general form: can computer programs make moral judgments? Imagine that programs can, indeed, make moral decisions. Imagine that such a program is better at that than the average citizen. Would that mean that we can compel everyone to use such a computer program? That we can do away with democracy and the constitutional state and put the development of our society in the hands of a few computer programs? Doing so is obviously out of the question. But in the context of a possible *merger* of humanity and technology—we refer to the view of Kurzweil as described in chapter 6—it is important to explore these questions.

In this chapter, we will examine whether computer programs can make moral judgments: do they understand what morality is?[1] *The Matrix* addresses a number of ethical questions. We will concentrate on this one question in this chapter and put it in a larger framework in the next. We will start with an exploration in *The Matrix*. Then we will carry out some tests to see if programs are capable of making moral decisions. Finally, we will discuss whether we should delegate the responsibility of moral judgment to a computer program.[2]

2. Distinguishing humans from computer programs[3]

The human characters in *The Matrix* can be subdivided into different groups. First of all, the "real" people who are held captive by the machines in large "vats" without their knowledge. Their bodies are real, but they live in a virtual world; their experiences are signals the matrix sends to their brains.

The second category consists of the "liberated" people, like Morpheus, Niobe, Trinity, Neo, and Cypher. They used to float in "vats" as well. Their world was the virtual world of the matrix. But they were liberated by the inhabitants of Zion. Liberated people in the film can be recognized by the plug-in devices they have in the back of their head, down their spine, and on their chest and arms. These outlets suggest that their bodies—most likely their brains—contain hardware and software that form an interface for the matrix program. That means they are actually "cyborgs": a combination of humanity and technology. On the one hand, they are thoroughly human: they eat and drink, work and sleep; they are aware of what is good and evil, they have a sense of justice and they declare their love to each other. On the other hand, they are also partly machine: they can be plugged into the matrix and download all kinds of knowledge and skills.[4]

The third group comprises "real" people who are the natural descendants of the first inhabitants of Zion: "One hundred percent old-fashioned home-grown humans." Tank, Dozer, and Link come to mind, but also most inhabitants of Zion belong to this category. They have no outlets in their bodies, and they cannot be jacked in to the matrix.

---

1   In this book, we understand ethics to be a disciplined reflection on morality. However, we will, following everyday usage, sometimes use these terms interchangeably.
2   See also Julia Driver, "Artificial Ethics," in Christopher Grau (ed.), *Philosophers Explore* The Matrix (Oxford: Oxford University Press, 2005) 208–212.
3   Based on Matt Lawrence, *Like a Splinter in Your Mind: The philosophy behind* The Matrix (Oxford: Blackwell Publishing, 2004) 11.
4   Cf. Henk G. Geertsema, "Cyborg: Myth or reality?" *Zygon* 41 (2006) 2, 289–328.

In the fourth category we find "humans" who turn out to be "sentient programs." We can identify some programs thanks to their superhuman capacities, or because they have access to the secrets of the matrix. Among them are the Agents defending the system, the Architect who designed the matrix, the Merovingian, the Trainman who smuggles programs into and out of the matrix, and the Keymaker who helps Neo to reach the Source. Other programs reveal themselves as such, like the little girl Sati who is trying to escape from the machine world to keep from being deleted. It is soon clear that her parents Rama Kandra and Kamala, too, are programs. Finally there are programs that are slowly revealed to be computer programs rather than "real" people. The biggest surprise in that respect is the Oracle and her protector Seraph. This surprise renders the whole story rather uncertain: are they programs controlled by the machines, or are they illegal programs with a distinctive design and purpose?

3. As artificial as *The Matrix* itself?

Let us examine the fourth category—"humans" that turn out to be sentient programs—more closely. When watching the trilogy, you find huge differences between the programs not only in relation to their functions and capacities but also in terms of character and morality.

The Agents Smith, Brown, and Jones are cold-hearted guys; dressed in dark suits, they come across as dedicated businessmen, and they rely entirely on the system. When Morpheus tells Neo that the Agents are nothing more than sentient programs, we can recognize that. We do not feel that they have a real personality and we get the impression they do not have emotions. Only when Morpheus cross-examines Agent Smith in *The Matrix* does it seem for a moment that he acts out of "character," as is the case in the final confrontation with Neo in *The Matrix Revolutions*. We have difficulty identifying with the Agents. When Sati, the little girl that should be deleted, says that Smith is a bad man, we agree with her. We actually feel that all Agents are bad because they maintain the suppressive system of the matrix even though we ascribe their "badness" to their being programmed that way.

The Oracle creates a completely different image. She is a sympathetic woman—a bit of a mystery, but what Oracle isn't?—whose whole appearance suggests that she is a personality that cares for the destiny of humankind. Later in the film we discover that she, too, is a program. That is quite a shock for us as viewers. Can she still be trusted? Is she part of the system? Is she on the side of the Agents? Is she controlled by the machines? These questions keep haunting you throughout the trilogy. In the course of the third part, the conviction grows that she is on the good side. And at the very end—when the Oracle "confesses" that she supported Neo because she believed in him—only then are you entirely sure.

In Agent Smith's view, there is no morality. He claims that human values are as artificial as the matrix itself. This idea returns several times in the film. Rama Kandra, for example, states that the meaning of words in a program—in this case the four-letter word "love"—depends on the number of connections that that word has. As such, he implies that human values are nothing but constructs.

4. "This is the Construct": First test

How can you test if a program understands the difference between good and evil and is capable of making moral judgments? Such a test could be carried out as follows.[5] A researcher is seated in a room with a hall that connects two other rooms. One of the rooms is occupied by a human participant, the other one by a computer where the program in question is running. The researcher types a moral question at the terminal, for example: is a married man or woman allowed to commit adultery when he or she feels like having sex with another person? The answer appears on the researcher's screen. Suppose the answer given is that it is allowed, provided that the person's spouse does not find out. The researcher decides to ask further questions about the phrase "provided that." After all, does the morality of the act depend on whether or not the partner finds out? The answer that subsequently appears on the screen says that this "complication" is definitely important, as "honesty" about such things can be threatening for the marriage. This answer also prompts the researcher to ask for further explanation, and so the discussion continues. The crux of the test is the fact that the researcher does not know who he is communicating with. It could be the human participant, but it could just as well be the computer. If the researcher cannot work out who is giving the answers, then the logical conclusion would be that computers can make moral judgments.

There is a difficulty, however. Suppose that the researcher and the human participant are (neo)liberal and the computer is an orthodox Christian. In that case, the researcher could be misled by the differences in background between the human participant and the program. The researcher could easily draw the wrong conclusions because he cannot

---

5   In 1950, the British mathematician Alan Turing suggested a test that could be used to determine whether or not a computer is intelligent. We are using this test in an adapted form to determine whether or not a computer program can make moral judgments.

identify with the Christian's way of thinking (Christians reject adultery on moral grounds). But this difficulty can be simply resolved by programming the program's morality as (neo)liberal or by employing a researcher familiar with both traditions.

5. "Is this another training program?"—A thought experiment

What exactly does this test prove? At any rate, it shows that a program can make moral judgments that cannot be distinguished from those of a human. But does it mean that a program *understands* the meaning of good and evil? We want to answer these questions with another thought experiment.[6]

Suppose an American who does not speak Chinese is locked up in a room. The room is stacked with books containing Chinese characters. There is also an English instruction manual explaining how the characters should be edited. The Chinese, who are outside the room, can ask him moral questions by writing them on a piece of paper and putting the paper in a drop box where he collects the questions. The American can look up the characters in the Chinese books with the help of the instruction manual, write down the corresponding characters and compose the converted characters into an answer. Then he can put the answer back in the drop box for the Chinese to read. In this way, the American can debate with the Chinese on all kinds of moral issues related to their culture.

The question is whether the American can actually give a moral judgment on the moral issues put forward by the Chinese. Based on the first test, the question should be answered affirmatively. After all, the Chinese are pleased with the quality of the moral answers. But does that mean that the American *understands* the Chinese morality? Absolutely not. All the American really does is to edit the Chinese symbols. This procedure results in new symbols that go back into the drop box. The conclusion of this experiment is that a computer program that passes the first test is capable of editing symbols but does not understand or possess morality. The man in the room simply does not know what he answers—he merely combines corresponding symbols.

One could object that the problem in this experiment is that the American does not speak Chinese. But, suppose that he knows the Chi-

---

6 This thought experiment is based on the work of philosopher John Searle. Searle developed the so-called "Chinese Room" experiment to demonstrate that if a computer passes the Turing test, it still needs to be established that it is in fact intelligent. In this section we draw on a variation on his thought experiment to show that a program that passes the first test, not necessarily understands what morals are, or possesses morality.

nese language but not the Chinese culture. And suppose he is asked to what extent a Chinese woman should obey her mother-in-law. The American approaches the question in the same way: he follows the instruction manual, edits the symbols and composes an answer. The Chinese are happy with the moral advice, as in their view, the American is making a moral judgment. Returning to the question whether the American understands Chinese morality, the answer will again be "no." The only thing he does is edit the symbols. And although he understands the symbols themselves, editing them does not require that he understands the Chinese morals. In fact, the American may be able to read the answer—that states that a married woman should obey her mother-in-law—but he does not actually understand it or agree with it, as he does not know the Chinese culture. Even in this variant the conclusion remains that the American is capable of editing the symbols, but he still does not understand Chinese morality.

Several objections can be raised against this thought experiment. You could say that all factors (the American, the instruction manual, and the books with Chinese characters) understand Chinese morality. This assertion is interesting, since hypothetically the whole thing can be inserted into a computer program and be incorporated in a robot. Examples of this can be found in science fiction films like *2001: A Space Odyssey*, *Bicentennial Man*, or *I, Robot*. But this argument is also problematic, as instruction manuals and books still do not "understand" morality, so it remains unclear why "the things as a whole" would.[7]

We thus reach the conclusion that the second test demonstrates that even if a computer program does pass the first test—meaning that a program can make a moral judgment that is indistinguishable from that of a human—we cannot simply conclude that a computer program understands good and evil and possesses morality. Therefore, we should examine these questions from a different perspective.

6. A view from the inside

Philosopher Mark Rowlands says that such questions are best approached by confronting "a view from the inside" with "a view from the outside."[8] Philosophically speaking this distinction is debatable, but let us follow his suggestion and find out what the analysis from the inside produces.

What exactly does a computer "do"? How does a program "work"?

---

7 See also Lawrence 2004, 50 ff., discusses several objections against Searle's thought experiment.

8 Mark Rowlands, *The Philosopher at the End of the Universe: Philosophy explained through science fiction films* (London: Ebury, 2003) 233.

We have grown so used to computers that we rarely pose such questions. However, it is important to take a closer look at this. The thought experiment in the previous section provides a simplified depiction of how computers and their programs work. Symbols are entered via a drop box (the keyboard) and are then edited with the aid of an instruction manual (the program). The data used during the editing process are retrieved from the books with Chinese characters (memory). Finally, an answer is construed based on the instruction manual (program) that is then sent back out via the drop box (the screen). Based on this simplified analysis, we can distinguish three core steps in the computer process:

a. The different symbols like letters, words, numbers, and other symbols that are entered into the computer via the keyboard are converted into electrical signals;
b. These electrical signals are edited in the way prescribed by the program. If necessary, information ("ones" and "zeros") stored in the computer's memory is retrieved;
c. The editing process results in an *output* signal that is converted into symbols.

Looking at the process from the inside out, we only find electrical signals; there is no trace of moral reflection. If we closely examine these signals and see how the editing process takes place, we find that they are primarily subject to certain physical laws. But, processed signals never turn into moral/normed discernment. When studying the memory chips, we only find electrical connections, strings representing ones and zeros. We do not find moral knowledge. In short, no matter how we analyze a computer's hardware and software from the inside out, we will never find moral reflections, moral norms, or moral knowledge.

Our preliminary conclusion is therefore that the claim that a computer program understands moral considerations and possesses morality should be rejected. When in the film *2001: A Space Odyssey* supercomputer HAL discovers that astronaut Dave is planning to pull the plug, he says: "What are you doing, Dave? I think I am entitled to an answer." But in accordance with our analysis, we can now say that Dave does not owe the computer anything. HAL is executing a speech subroutine, but his words do not actually contain a moral judgment.

We emphatically use the caveat "preliminary conclusion." Suppose that humans turn out to be nothing more than a huge quantity of electrical signals (as many scientists suggest, see chapter 10), then it might not really be problematic to deny a computer program morality. In chapter

12 we will examine this issue in greater detail.

7. A view from the outside

Let's explore the workings of a computer program from a different angle, taking a view from the outside. When looking from the outside, the first thing you notice is that the program itself does not do anything. It all starts with the keyboarded act of a human being, a user who engages a moral question, one for whom that question is important. For some reason or another, he or she wants an answer. The answer that appears on screen is again interpreted by a human. He or she does not see just symbols but reads a message that has meaning. The human may or may not accept that message, and he or she may or may not act accordingly.

If you look at a computer program from the outside, you can see that the program reacts—you might call that a moral judgment—but that this "response" is nonetheless evaluated (morally) by a human being. He or she will have to decide what that "judgment" implies and whether or not it is a sound judgment. In other words, it seems that the question as to whether a computer understands the difference between right and wrong can only be asked in relation to a human being.

What does this analysis show? In our view, the most important conclusion is that such a computer program functions as an "object" of an acting human. In other words, the acting human takes a central position, and that acting human uses technical artifacts or constructs. One such artifact is a computer program, or, in this case, the *output* of such a program.

We will illustrate the conclusion that computer programs function as objects of the acting human with two examples. First, computer programs are used a great deal in the financial world. In everyday speech, we say that the computer sells and buys, but we all know better. Also at the stock exchange the computer's programming is an object that is subject to human actions. Even if a computer can decide "independently" on the purchase and sale of the shares, its programming is always based on the economic judgments of humans that have been built into the program as preconditions.

The second example is related to the use of computers in the court system. Even if *computer-aided administration of justice* was applied, an analysis from the outside shows it is absurd to suggest that a computer can make legal judgments. The computer remains an object in human hands at all times.

The view from the outside shows that humans understand what is right and wrong, that humans operate economically, that humans administer justice. In other words, humans act as responsible subjects in moral, economical, and legal acts. Human beings are conscious of moral norms, have knowledge of economic systems, and possess a sense of law and justice. The Christian philosopher Nicholas Wolterstorff argues that that is why animals (and in our case, machines) lack the power of *speech*. Because speech requires the ability to take normative positions with regard to others; for example, making a promise means that you owe the other person something.[9] One will be called to account for one's actions. This will not happen to an animal or a machine. Based on the view from the outside, we can see that the computer has a different "role." It only functions as an artifact or passive object in the acts of humans. The computer "does" things, but those things are used by humans. The analysis from the outside thereby demonstrates the great difference between the moral consciousness of the acting human and the computer aiding the acting human in the process; in these acts, humans act as responsible subjects, while the computer is the passive object.

It is exciting to compare this analysis with the analysis from the inside, which focuses on computers performing physical "acts," like processing symbols or electrical signals. If we extend the meaning of the word "subject" a little, we could say that the computer does function as subject in these actions.

The analysis above can be extended to a number of other examples and thought experiments. We can conclude then that, in the terms used in Reformational philosophy, a computer does function as a subject in several cases, e.g., in the mathematical and physical aspect (cf. Chapter 3). Furthermore, we conclude that humans function as subject in many ways, first of all in the mathematical and physical aspect, in addition to in the social, economic, juridical, ethical, and faith aspect. But in the very situations where critical/evaluative actions are concerned, such as in the social, economic, juridical, ethical, and faith dimensions of life, we see that the computer functions "merely" as *object*.

Hence, we conclude that neither the "view from the outside" nor the "view from the inside" provide sufficient arguments to prove that computer programs know what is good and evil. In fact, there seems to be a fundamental difference between the human being functioning as a responsible subject in the moral aspect and the computer functioning merely as a passive object in this aspect. However, this conclusion triggers another question: suppose that a computer program functions so well

---

9   Nicholas P. Wolterstorff, *Divine Discourse: Philosophical reflections on the claim that God speaks* (Cambridge: Cambridge University Press, 1995) 82–85.

that it passes the first test (see section 4)—should we not just allow it to be applied to moral issues?

8. God's point of view

The Matrix paints a challenging scenario. If people can be provided with "bio-ports," if there are computer programs that pass the first test with flying colors, and if we can download such programs, would it not be a giant step in the development of a righteous society? Wouldn't we be able to banish crime from our society and greatly improve its moral standards—in a much better and much more effective way than in the film *Minority Report* (see Chapter 10)?

Philosopher Joseph Weizenbaum extensively studied the possibilities of information processing techniques.[10] His primary aim was not to see what computer programs would be capable of doing but to explore what these programs would be *allowed* to do. In his view, there are some applications that you simply should not want and others that you should want to apply only with the utmost caution. He even refers to the first type of applications as "obscene." These are applications where the mere consideration "ought to give rise to feelings of disgust in every civilized person."[11] He argues that all applications where a computer program is substituted for "a human function that involves interpersonal respect, understanding, and love" are morally reprehensible; these are functions where a computer program "*ought* not to be substituted."[12]

Weizenbaum passes a moral judgment on the application of computer programs in areas where human functions like respect, care, justice, love, and faith play an important role. In his view, programs *ought* not to be applied in these matters. His view is as experienced as it is sympathetic. Everyone intuitively feels that raising children cannot be done by computers, and everyone intuitively feels that giving love and care to your wife or husband should not be done via a computer. But are these objections sufficient? Suppose that our arguments that computer programs have no sense of good and evil would bear the scrutiny of criticism. Suppose that everyone acknowledges that these programs do not have morality. Even still, the question arises why a program that passes our first test should not be allowed to be used, especially when it turns out that such a program passes better judgments than the average American.

---

10 Joseph Weizenbaum, *Computer Power and Human Reason: From judgment to calculation* (San Francisco: W. H. Freeman, 1976).
11 Weizenbaum 1976, 268–9.
12 Weizenbaum 1976, 269–270.

Isn't it tempting to apply such a program?

It is obvious that Weizenbaum would not be impressed by this latter argument, as he takes a different perspective on things. Eventually, his key concern is human responsibility. And starting from this responsibility he addresses the matter of the applications of computer programs. We share Weizenbaum's view. We are convinced that we should discuss such questions from the perspective of human responsibility. Only from that perspective can we decide whether or not we should apply a certain program; meaning that the question of a program's effectiveness—for example, how well a computer does in comparison with a human being—is preceded by the question of human responsibility.

In our view, people are responsible beings. We are responsible for our deeds, and will also be held accountable for these deeds: to fellow human beings, to a judge (in trying situations), and eventually always to God. This responsibility cannot be delegated. Seen from that perspective, a computer's programming is and remains an instrument in the hands of humans. Instruments ought never to operate autonomously and independently of humans.

Mark Rowlands demonstrates that some questions cannot be answered from "a view from the inside" or "a view from the outside." He refers to these as "strange" questions, pushing us to "the limits of reason."[13] The question of human responsibility ultimately cannot be answered by either a view from the inside or a view from the outside. These "strange" questions require not that reason be suspended but that the debate be held at a different level, where we can discuss the nature of morality, the origin of good and evil, and the meaning of these for human existence. The question of human responsibility is therefore in our view also a question of our origin. Our deepest convictions dictate—and we stand in the long line of the reformed tradition—that the question of humanity's origin leads to God the Creator. Only from the perspective of the Creator—from God's point of view—can we explain the most vital elements of human responsibility.

The Bible's perspective on humankind is clear: male and female human beings are created as responsible beings. They must answer God's call to cultivate the earth. They may not delegate that responsibility to technological artifacts or programs that can make moral judgments. Even as aids in the hands of humans, we should always apply them in wisdom, knowing that human responsibility remains the basis of principle. This will be discussed in the following chapter.

---

13 Rowlands 2003, 176.

Chapter 12

MORALITY IN A HIGH-TECH SOCIETY

> Trinity: "You can't be dead, Neo, you can't be because I love you.
> You hear me? I love you!"
> (*The Matrix*)

> Morpheus (about the agents of the system):
> "Their strength and speed are still based in a world built on rules."
> (*The Matrix*)

> Smith: "Without purpose we would not exist.
> It is purpose that created us. . . .
> It is purpose that defines."
> (*The Matrix Reloaded*)

1. "I protect that which matters most"

Science fiction films are full of moral questions, as we have seen in the previous chapters. On the one hand, this is manifest in the dialogues and discussions about how to behave in a certain situation. On the other hand, we find it in the development of technology. In this chapter we will put these moral questions in a broader framework.

One of the most interesting dialogues about the purpose of human action is between Agent Smith and Neo during their final trial of strength in *The Matrix Revolutions*. In this dialogue, Agent Smith asks Neo why he keeps fighting. Does he think he is fighting for something? For something more than survival? For freedom? For truth? Perhaps it is for peace? Or could it be for love? In Smith's view, all human values that serve as a basis for our moral judgments are "constructs" of a feeble human intellect. They are as artificial as the matrix itself. From Smith's point of view, it is useless that Neo keeps fighting. He rejects the idea that there could be "valid reasons" for choosing a particular way. He does not accept any moral value. And, by extension, he accepts no moral claim whatsoever.

How should we view Agent Smith's position? There are two options: either he is a moral nihilist who rejects every moral judgment or he is a moral subjectivist who believes that all moral judgments depend on the preferences of the individual. In this situation, these options are two sides of the same coin. Both positions start from the assumption that there is no "universal truth" on which moral claims can be based. It is therefore useless to make moral claims.

Yet, there is more to say about Agent Smith. He discovered that the only goal in life is "to end." He professes the instrumental morality of the machine society: each program has a goal, and each appliance has to perform a certain function. After some time, the program is no longer needed, and the appliance is outdated. The program can be deleted, and the appliance can be destroyed. In the light of this "discovery," Smith's position can be best characterized as that of a moral nihilist: he does not acknowledge human values. In his view, values such as freedom, truth, peace, and love are unrelated to human society. He manifests himself as an unscrupulous, all-destroying, self-destructive *Wille zur Macht* (Nietzsche).[1]

Is Smith's view true? Are moral judgments nothing but human constructs? Is the instrumental goal or technical purpose the only thing that matters? Are there no universal values, such as freedom, truth, peace, or love, on which we can base our ethics?[2] We will start with an exploration of morality in *The Matrix* and *Star Trek*. Then, we will examine three classical philosophical ethical theories: virtue ethics, deontological ethics, and teleological ethics. We will conclude with a Reformational philosophical view on responsibility.

2. And if morality isn't relative?[3]

Do universal moral norms exist? You could object that the small group of rebels in Zion ought to be considered as a monoculture, one with its own values and norms. If several cultures had survived on earth after the wars had ended, then we would have seen all kinds of differences between them. In that case, we would have to acknowledge that there are no universal moral values but that each culture has its own moral code. Philosophically speaking, we can characterize this position as moral relativism. We want to explore this matter by means of the science fiction

---

1 See also Mark A. Wrathall, "The Purpose of Life is To End: Schopenhauerian pessimism, nihilism, and Nietzschean will to power," in William Irwin (ed.), *More Matrix and Philosophy* (Chicago: Open Court, 2005) 50ff.

2 Again, ethics can be seen as the (scientific) reflection on moral, although we will not thoroughly avoid using these terms mutually exclusive as in everyday speech.

3 After a paragraph title by Judith Barad, Ed Robertson, *The Ethics of Star Trek* (New York: HarperCollins, 2000) 23.

series *Star Trek*, where we find many different peoples and races (see also chapter 7).

Star Trek discusses a number of dilemmas set against the backdrop of a fictional future where a spaceship explores planets of other stars in the Galaxy. During this exploration, they meet all sorts of races: the intelligent Vulcans for whom logic is the basis for every decision, the warlike Romulans who have never learned to control their emotions, the fighting Klingons who put honor above all other things, the Borg who have enhanced themselves by means of implants and always act collectively, and the Ferengi with their mercantile obsession with profit and their patriarchal society.

Star Trek raises a large number of moral questions. In the first place, we can find moral questions and dilemmas in relation to rules, respect, power, justice, and responsibility. There are questions and dilemmas that still play a role in a society technologically much more advanced than ours. In the second place, we can find all kinds of new questions concerning the use of new technologies and the interaction between entirely different cultures. The first thought could be that given its cultural pluralism *Star Trek* would advocate a relativistic orientation. One would also assume that universal cultural values would have no credence, especially because the most important guiding principle since the foundation of the Federation forbids any interference with the natural development of other civilizations (the *Prime Directive*).

This is not the case, however. Judith Barad extensively studied the ethics in this series, concluding that *Star Trek* shows that "any culture can be held up to universal ethical standards."[4] The series' pluralism is based on certain universal moral values. Genocide, tyranny, slavery, racism, and exploitation are rejected. In addition, the Federation forces cultures that promote sexism, that applaud terroristic attacks, that lack attention for the poor and homeless, and that dump toxic waste in space, to reconsider their actions. In short, cultural relativism is being rejected.

It is not surprising that universal, moral norms can be found below the surface of the events. We can see that *Star Trek* reflects an American picture of the world in 1960–2000—written with an eye to the ideal of sexual and racial equality. Men and women, white and black, inhabitants of different planets search the universe together for new civilizations. In various stories, racial discrimination, the Vietnam War, and the Cold War are implicitly criticized.[5] The message that Gene Roddenberry, the

---

4   Barad and Robertson 2000, 23.
5   Barad and Robertson 2000, 22.

creator of the first series, wanted to bring across to Americans was that different sexes and races are equal and capable of working together.

The conclusion is that important science fiction films like *Star Trek* presuppose a number of universal moral values that ought to direct human actions in a high-tech society.[6]

3. "You hear me? I love you!"

In the previous section, we asked whether or not universal moral norms exist. We have seen that science fiction films presuppose a number of moral norms that apply to all peoples and cultures, and we want to continue in this vein. Philosophical ethics distinguishes three main approaches that presuppose the existence of universal norms: virtue ethics, deontological ethics, and teleological ethics.[7] We will discuss these one by one to decide whether they can (help to) form the basis of a moral code in a *high-tech* society. We will start with virtue ethics.

Virtue ethics focuses on the moral qualities and lifestyles of an individual wishing to achieve a well-lived life. A key element is the personal capacities of the acting person; that is, the ability and skills to reach a certain goal. Greek thinkers thought of virtues like wisdom, courage, and justice; typically virtues that were important for life in the Greek city-state. Medieval thinkers distinguished seven virtues, including the Christian *triad* of faith, hope, and love.

The name of Aristotle (384–322 BC) is closely linked to virtue ethics. In his book *Nicomachean Ethics*, he gives a beautiful description of the virtuous life. In his view, the ultimate goal of human actions is striving for the highest good: *eudaimonia*, or the proper life, wellbeing, or happiness. In daily life, words like "good" and "happiness" are often linked to subjective experience: something feels good, or you feel happy. But, in Aristotelian thought, the word "eudaimonia" has a different meaning: it refers to a visibly successful life. It refers both to a concrete lifestyle and concrete actions (virtuous life) and to the goal or *telos* of human actions (a good or proper life). In other words, when you are good at something, it helps you to reach wonderful goals. A good example in *The Matrix* is Niobe's outstanding ability when it comes to driving a hovercraft. She uses this ability so well in the war against the machines. Action and goal are thus intrinsically linked in the concept of "eudaimonia." In the words of Alasdair MacIntyre, it is not just about *behaving well* but also about

---

6 Gregory Basham, "The Religion of *The Matrix* and the Problem of Pluralism," in William Irwin (ed.), *The Matrix and Philosophy* (Chicago: Open Court, 2002) 11 ff.

7 Because we do not discuss the question of what universal norms are exactly, we put the adjective "universal" in quotes.

faring well, it includes both *well-doing* and *well-being.*⁸

The highest good can only be achieved by leading a virtuous life. The Greek word for virtue is "arete," meaning both "virtue" and "excellence." A virtue is a character trait that must be developed through proper training and ample practice. It is only possible to become a brave soldier when you act as a brave soldier in battle. It is only possible to become an honorable engineer by acting honorably when designing new products.

Which virtues are important? Answering this question actually precedes the virtue ethics approach. Aristotle emphasized virtues that were important in the Greek city-state, while the church fathers pointed to the virtues derived from the Bible. Many modern technocrats focus on virtues that contribute to the development of technology and the economy.

Aristotle's most significant contribution to ethics is that he connects the good to daily life in a direct manner. In his thought, the good is *intrinsically* related to a certain life-practice, and his ethics can be effectively applied to all sorts of ethical questions in our *high-tech* society. It is no wonder that *Star Trek,* to a large extent, reflects the virtue ethics perspective. The captains of the spaceships are depicted as virtuous men characterized by virtues like courage, friendship, justice, and peacefulness. Similarly, the main characters in *The Matrix*, Neo and Trinity, clearly display the virtues of courage, perseverance, love, and loyalty: "You hear me? I love you!"

In *Star Trek,* Captain Jean-Luc Picard and his crew visit the planet Rubicam III in the Strnad system.⁹ The inhabitants of this planet, the Edo, are extremely kind, and they decide to stay a little longer. During their stay, they discover that the Edo society is controlled in a *Big Brother*-like way. Various life-zones are observed closely via technical systems. Breaking the law is punished by death. The population does not know which zone is observed when. When the young Wesley is sentenced to death for accidently breaking the law, Captain Picard is faced with the choice to follow the Federation's *Prime Directive* (that forbids interference with the customs of other civilizations) or to break it. Picard decides to do everything in his power to save Wesley's life. He pleads for Wesley's life with the argument: "There can be no justice as long as laws are absolute."¹⁰ In Aristotle's view, Picard makes the right choice because he demonstrates what justice means in a concrete situation because he shows that justice is more than just what is written in the law. Picard possesses

---

8   Alasdair MacIntyre, *A Short History of Ethics* (London: Routledge, 1998)², 57.
9   Barad and Robertson 2000, 119 ff.
10 Barad and Robertson 2000, 124.

the virtue of justice because in this specific situation he acts in such a way that he realizes a just goal. In another episode, Picard says that he violated the *Prime Directive*—despite his sworn obedience—on more than one occasion ". . . because I thought it was the right thing to do."[11]

Star Trek and *The Matrix* clearly show that a high-tech society also requires virtuous actions. The development and use of technology ought to be sustained by virtues like justice, integrity, love, and loyalty. We conclude that virtue ethics can help form a basis for an ethics for a high-tech society, but it does need a framework that prescribes which virtues are important.

4.   "…based on a world built on rules"

Deontological ethics offers an entirely different approach. Can it contribute to a bona-fide ethics for a high-tech society? Deontology is not primarily about the good life but about the question whether or not an action agrees with certain moral rules (laws, norms, principles). The focal point is the moral quality of the action per se. After all, a system of commands can help to remediate a lack of virtues. One of the most famous contributors to the development of deontological ethics is the German philosopher Immanuel Kant (1724–1804).

Kant maintained that one's actions should not be dependent on values and norms outside oneself, as is the case in virtue ethics, for example. He argues that one's moral decisions should be based on the (autonomous) ability to make rational decisions for oneself: "Have courage to use your own understanding." Based on this principle, Kant formulated his famous concept of the categorical imperative in his works *Grundlegung zur Metaphysik der Sitten* (1785) and *Kritik der praktische Vernunft* (1788): "Act only in accordance with that maxim where you can at the same time will that it becomes a universal law." An imperative is a command: you are compelled to act this way. A categorical imperative is an unconditional command: you are compelled to act this way, regardless of the consequences of your actions in a specific situation. The essence of the categorical imperative is that you act in such a way that your action can be generalized to a universal law. Some things simply must be done (or are simply not allowed), end of discussion.

Kant gives the example of a promise you cannot keep. Is it morally correct to make a promise you cannot keep? The categorical imperative states this is morally acceptable if I can want that everyone is allowed to break a promise. But, the point of making a promise is that you will keep

---

11   Barad and Robertson 2000, 125.

it. If breaking a promise becomes a universal law, then it is pointless to trust that others will keep their promises. It results in logical contradiction, as I cannot simultaneously want a promise to be kept and to be broken. The conclusion is that you may not make a promise you cannot keep. In other words, it is your duty to always keep your promises. In Kant's view, an action is thus morally acceptable only if following that maxim results in an acceptable situation.

In nearly every episode of *Star Trek* ethical choices are made based on the idea that someone has to perform his duty. This is manifested in particular in the ethical dilemmas that the captains of spaceships are facing. In the episode "I, Borg," for example, Captain Picard has to choose between his desire to take revenge and to fulfill his moral duty.[12] In this episode, the crew of the Enterprise searches the wreck of a small spaceship and finds one survivor: a seriously injured young Borg. The Borg are a dangerous humanoid species that have acquired enormous (war) power by placing all sorts of implants into the human body. Picard proposes to equip the young Borg with a virus program where the entire Borg race— that often take an aggressive attitude towards the Federation—could be destroyed. However, the young Borg recovers as a result of the good medical care and increasingly displays human character traits. Picard considers it his moral duty to protect the young Borg's individuality, but he also wants to use this opportunity to eliminate the dangerous Borg. What should he do?

In *The Matrix* we also find several characters who clearly act from a sense of obligation. Agents Smith, Jones, and Brown are the first to come to mind (at least in the first film). Morpheus says their power and speed is "based on a world built on rules." Another example is Commander Lock, who was responsible for the defense of Zion.

Kant also gave a more practical formulation of the categorical imperative. This formulation is of great significance for the use of technology in relation to humanity. It stresses that humans may never be used solely to attain your own goals. He says: "Act in such a way that you treat humanity, whether in your own person, or in the person of another, always at the same time as an end and never simply as a means." Never treat another person simply as a means: that was the problem in *The Matrix*. The machines used human bodies as an energy source for themselves. Never treat another person simply as a means: that was the problem in the science fiction film *The Island*, where human clones are fabricated for transplantation purposes. Never treat another person simply as a means:

---

12 See Barad and Robertson 2000, 207 ff.

that is the problem in our high-tech society where people are used as cogs in the machine of technical organizations and where clients are a means to generate profit.

Kant's deontological ethics are often found in technical organizations. All standards related to equipment, products, processes, and systems can be traced back to deontological ethics. All guidelines concerning the use of equipment, all safety requirements for machines, all privacy requirements used in information technology—they are all Kantian in fashion. The same is true for the codes of conduct for engineers. These regulations possess a certain power; that is the beauty of Kant's deontological ethics.

One of the most significant problems in Kant's ethics is that he fails to provide a solution for the situation where two duties are in conflict. Another problem is that he pays insufficient attention to the negative effects that acting according to a certain norm may have in a concrete situation. Both problems occur in the situation Jean-Luc Picard is confronted with by the Edo people (see previous paragraph). From the perspective of the *Prime Directive*—the maxim would be "do not interfere with the customs of another civilization"—he would not be allowed to intervene. But from the perspective of acting rightfully he would have reached a different conclusion. In that case—the maxim would be "interfere with the customs of other civilizations if they act unjustly"—he would be obliged to intervene. This situation generates conflicting duties. The second problem surfaces as well; it was clear to Picard that the effects of non-intervention (the death penalty) were unacceptable.

5. "It is purpose that created us"

In the last two sections, we have seen that the effects of an action are indirectly taken into account in the moral judgment. Virtue ethics centers on the virtues that are necessary for achieving the good. In the discussion of deontological ethics, it appeared that doing one's duty sometimes unjustly disregards the effects. There is also a philosophical approach that emphasizes the goal or consequences of the action on a certain situation. It is referred to as teleological ethics, or consequentialism.

In teleological ethics, the focus is on the result of the action. It centers on the question of whether the goal we strive for is good or evil. Usually, this goal is described in terms of advancing human happiness by maximizing pleasure and joy and minimizing pain and suffering. Classical utilitarianism speaks of the "greatest happiness principle": the greatest good for the greatest number of people. In contrast with virtue ethics, teleological ethics sees no intrinsic relationship between the action and the desired goal. Teleological ethics is distinguished from deontological ethics in that the former pays no attention to the moral quality of the action in itself. Two actions that have the same result are considered morally equal.

The most famous philosophers that shaped consequentialism—more precisely a variety called "utilitarianism"—are Jeremy Bentham (1748–1832) and John Stuart Mill (1806–1873). Bentham in particular tried to develop *the greatest happiness principle* as much as possible in a quantitative manner. He held the view that the "positive" and "negative" effects of an action, like pleasure and grief, can be measured, added, and subtracted, so we can decide whether that action is good or bad for society. In his time, Bentham was a revolutionary. He actively contributed to the solution of a variety of social issues such as the reformation of criminal law. The *greatest happiness principle* provides a guideline for determining the punishment scale for a certain crime. And in his calculations, the happiness or grief of a tramp was given the same value as that of a count!

Consequentialist theory has a number of problems. It is not at all self-evident that positive and negative effects can be added and subtracted. Besides, applying only one criterion (happiness or pleasure) seems rather meager. There are many other values that cannot be reduced to this criterion, such as environmental preservation. This approach presupposes a certain choice of values that precedes the application of this theory! And it is not entirely true that only the consequences should count in making a moral judgment. The end does not always justify the means. Sometimes something is simply wrong, even if the consequences are not so bad. Two actions that have the same result, such as hurting someone on purpose or by accident, are certainly morally different.

These problems do not mean that we should reject teleological ethics as a perspective. If we acknowledge that in a high-tech society, fundamental values such as love, peace, and justice are of great importance, then teleological ethics can contribute to posing the right questions concerning the development of new technologies. Could a certain product contribute to a more just society? Could a new technology promote peace on earth? Could a more complex system contribute to people's care for each other? We are under no illusions about these questions being easily answered, but teleological ethics does help in the process of asking relevant questions.

In *The Matrix*, the concept of "purpose" plays a crucial role. In the confrontation between Agent Smith and Neo in *The Matrix Reloaded*, the former declares to the latter: "…without purpose we would not exist.

It is purpose that created us (…) it is purpose that defines us." A similar thought is articulated by the so-called Keymaker. When Niobe asks him why he knows everything about how to access the Source, he answers, "I know because I must know. It's my purpose." And, obviously, the choice that the Architect offers Neo in *The Matrix Reloaded* is also one with two very different outcomes: the death of Trinity or of all other people.

In *Star Trek* we often find teleological reasoning. We give two examples: one "good" and one "bad" example. A crucial scene in the film *Star Trek II: The Wrath of Kahn* contains *the greatest happiness principle* in optima forma. When Spock (part Vulcan, part human) gives his life to save the spaceship *Enterprise*, he speaks the memorable words: "The needs of the many outweigh the needs of the few . . . or the one." He repeats this statement in slightly different words at the end of the film: "The good of the many outweighs the good of the few or the one." In light of Spock's background, this choice is understandable. He is part Vulcan, and the distinctive trait of the Vulcans is that they only make decisions based on rational arguments.[13]

In the film *Star Trek: Insurrection* a utilitarian dialogue takes place in order to judge a certain proposal on its moral merits. Admiral Dougherty wants to move the 600 inhabitants of the planet Ba'ku to another planet without their consent. The reason is that this planet has a number of rings that have healing power that can cure millions of people. Admiral Dougherty tries to persuade Captain Jean-Luc Picard of the *Enterprise* spaceship that the public interest prevails: "Jean-Luc, we are only moving 600 people." Picard believes that a consequentialist approach is unacceptable in this situation. He responds, "How many people does it take before it becomes wrong—a thousand? Fifty thousand? A million? How many will it take?" In addition to that, the action is in violation of the *Prime Directive*. Picard decides not to execute the admiral's orders; instead, he protects the interests of the Ba'ku and risks his own career.

The example of the Ba'ku shows that teleological arguments require a moral framework: *which* goals and consequences count, and how should they be weighed? Without a moral framework, teleological ethics gets reduced to a "rational" or "technical" debate. Particularly in the process of technological development—take the famous Pinto case for example[14]—we are in danger of fixating on the targets of effectiveness

---

13 Interestingly, in *Star Trek III* Captain Kirk as righteous virtue ethicist reverses Spock's expression—"the needs of one outweigh the needs of many"—as a reason why he risks the lives of his people to bring back Spock.

14 In 1968 Ford started developing a new car: the Ford Pinto. This car was supposed to be put on the market within two years. The car was not to weigh over 2,000

and efficiency. That is why it is crucially important to understand that the choice of values that shape the criteria for making decisions precedes teleological ethics.

6. "You're the one who has to walk through it": the responsible person

In the previous sections, we discussed three classical approaches in philosophical ethics: virtue ethics, deontology, and teleological ethics. We pointed out that each of these approaches offers a unique (valuable) perspective. We also noted in passing that each approach presupposes the existence of universal moral principles. In virtue ethics, these values are related to the virtues of the actor; in deontology they are related to the norms or rules for the action; and in teleological ethics they are related to the assessment of the consequences or the intended purpose.

In our opinion, responsible acting requires an account of responsibility from each of these perspectives. Not only is the morality of the actor concerned here but the morality of the action and the morality of the goal. In our view, these three perspectives can be integrated in the concept of responsibility. Responsible people formulate goals and act according to them. The human actor is responsible for one's fellow human beings, society, and the natural environment. This responsibility goes beyond thinking in terms of virtues, duties, and goals. Consider the challenges we are facing in our western society: the development of modern technology brings with it questions related to resource depletion, water and air pollution, waste management, and climate change. Other difficulties have to do with access to and the cost of technology, also as these relate to the economy in the first, second, and third world.

Change agents are clearly responsible for addressing the questions that are provoked by the development of technology (especially when this is done with an eye to "growing" the economy). They may never shy

---

pounds and not cost over $2,000. In 1978 it turned out that something was wrong with the design. In one accident the fuel tank ruptured and the Pinto caught fire. Thee passengers died. The cause was a dangerous design flaw. During the legal investigation, it appeared that Ford had been aware of this flaw from the beginning. The designers had indicated that the problem with the fuel tank could be solved with a relatively cheap adjustment in the design (11 dollars per car). Yet Ford did not make the change. A cost-benefit analysis had shown that redesigning the Pinto would result in a lower benefit to society than not adjusting it. In this calculation, the social costs of traffic casualties were estimated to amount to $200,000 per death and $67,000 per injury. Adjusting the design would add up to $137 million per year, and not adjusting it to 49 million. So they decided against adjusting the design.

away from this responsibility. Not doing anything is also a choice; even walking away is a way of choosing. But even when one accepts this sense of responsibility, the question remains as to how best to follow through on that commitment. Or to put the question more broadly: how ought a high-tech society to develop?

In discussing virtue ethics, deontology, and teleological ethics, we pointed out that none of these approaches provides a real answer as to which virtues, duties, or goals are morally sound. All of these approaches presuppose a certain view on morality. Aristotelian virtue ethics, for example, reflect the views of the Greek polis. Ultimately, the question as to which virtues, duties, or goals are morally sound cannot be (fully) answered by philosophical ethics. The answer to this question (among others) lies rooted in one's religion or worldview. One's (religious) convictions define one's sense of purpose, the place and value of one's neighbor, and the need to care for the ecology of one's environment. That is why it is so difficult to talk about "universal" values and norms, for the adjective "universal" suggests that there is a (rational) way of reaching agreement about what is morally sound. But such a (rational) way does not exist. On the contrary, we find that many people believe that the values and norms they glean from their religious convictions or worldview are of universal significance.

We believe that what defines one's priorities regarding which values and norms ought to be leading the development of our high-tech society is ultimately a matter of one's religious or ideological commitment. That is why it is not so surprising that religion plays such a key role in many science fiction films (in particular, the later series of *Star Trek*, *Star Wars*, and *The Matrix*). We firmly believe that human beings are called to respond to God the Creator. Which means that everything that is should be oriented towards him. That orientation implies that we can be convinced about the value of our own life, that we should be filled with love towards our neighbor, that we should aim for justice and peace in our society, and that we should watch over the ecology of our environs with care. It also means that we believe that the Bible's laws, commands, and principled instruction are good for the development of our society. Within that context we dare to use big words like love, care, justice, and peace—matters that the next chapter picks up on.

Let's return to the scene cited at the opening of this chapter. During the final great fight Agent Smith asked Neo, "Why keep fighting?" Let that question keep burning in our hearts. Why should one keep fighting for serviceable technology? Why keep fighting for a just society? The answer can only be: because we are responsible people.

Chapter 13

BELIEF AND RATIONALITY

> Trinity: "Morpheus believes he is the One."
> (*The Matrix*)

> Morpheus: "Some of you believe as I believe, some of you do not. Those of you that do, know we are nearing the end of our struggle. The prophecy will be fulfilled soon."

> The Merovingian: "You see, there is only one constant, one universal. It is the only real truth. Causality."
> Morpheus: "Everything begins with choice."
> (*The Matrix Reloaded*)

> Seraph: "Did you always know?"
> The Oracle: "Oh no. No I did not. But I believed, I believed."
> (*The Matrix Revolutions*)

1. "Everything begins with choice"

In Chapter 11, we categorized the different people we meet in *The Matrix*. Hence, we distinguished four types of people: (a) "real" people who are locked up in a tub, (b) "liberated" people like Morpheus, Neo, and Trinity, (c) "real" people who are the natural descendants of the first inhabitants of Zion, like Tank, Dozer, and others, and (d) "sentient" programs like the Oracle, Agent Smith, and the Merovingian.

We could also characterize the people in *The Matrix* in another way: based on where they fall on the continuum between belief and surrender on the one hand, and reason and control on the other hand. In that sense, the most remarkable person in *The Matrix* is Morpheus. In all three films, he is portrayed as a driven man: a man who is certain that the prophecy will be fulfilled, who is convinced that they will win the battle against the machines. In short, he is someone whose faith in the positive outcome is as firm as a rock. If you wonder where this faith comes from, there is really only one possible answer: *he believed.*

Morpheus's most notable opponent is probably the Merovingian. The Merovingian is a program that possesses the knowledge and power to access the Source (the location in the matrix where redundant or dysfunctional computer programs are deleted and new ones are made). He exerts his power through the Keymaker, among others. In addition, he is the "boss" of the mysterious *Trainman,* who can smuggle programs into and out of the matrix. The Merovingian is the archetype of rational control. In his view, causality, rational considerations, and technical control are everything.

It is easy to categorize the other main characters along the same lines of Morpheus versus the Merovingian. Neo, Trinity, Niobe, and the Oracle are all in the category "believers" and Agent Smith, the Architect, and the *Trainman* fall into the category "rationalists."

There are two scenes in the trilogy where the fight between these opposites reaches a climax. The first scene is found in *The Matrix Reloaded,* where Morpheus, Trinity, and Neo visit the Merovingian in order to find the Keymaker who can take them to the Source. At that point, a philosophical debate on freedom arises (see chapter 10). In that discussion, the Merovingian asserts that there is only one truth: that of causality. Morpheus does not agree with that and states that "everything begins with choice." The word "everything" is probably the most striking in this reply. First of all, Morpheus expresses that his confidence in the Oracle's prophecy and his faith in Neo as savior of humankind is a *choice*, a *faith choice*. But, at the same time, Morpheus indicates that the Merovingian's confession in relation to causality and determinism is also a *choice*, a *faith choice*. Morpheus demonstrates that "believers" and "rationalists" have much in common: they choose.

The fight scene between the opposites is the closing scene in *The Matrix Revolutions*; in chapters 10 and 12, we also referred to this crucial fight. Agent Smith asks Neo why he keeps fighting. Smith cannot think of any good reason, except for the crude fight for survival. In his view, every motivation, be it the fight for freedom, truth, peace or love, is a "construct" of the "feeble human intellect." According to him, the only purpose of life is to finish well. Neo does not enter into a rational debate with Smith. He gives the only reply possible in this situation: "I choose to." Here, too, we clearly see the idea that belief precedes rationality. An idea that is, to put it euphemistically, not generally accepted in philosophy.

In this chapter we will explore this theme in *The Matrix* trilogy, search for the roots of the theme in the work of the Danish philosopher

Søren Kierkegaard and conclude with our own Reformational philosophical view on belief and thinking.

2. "Load the jump program": leap of faith

The Matrix starts with a gripping scene where Trinity is chased by Agent Smith. She is pursued across the rooftops of tall buildings, they jump from one rooftop to the next, and in between a terrifying abyss is yawning. Scenes like this can be found throughout the trilogy. An important theme in *The Matrix* is overcoming the fear of falling and leaning to leap in faith. You can—as Morpheus teaches Neo during his "jump training"—only jump from one building to the next when you leave your fear, doubt, and disbelief behind.

Morpheus impressively demonstrates what a leap of faith really means. After visiting the Oracle, Morpheus is captured by Agent Smith. Neo and Trinity return to the matrix to free him. During this action they use a helicopter where Trinity sits in the cockpit while Neo hangs by a rope. The windows of the building where Morpheus is held have been smashed in the shooting. Amidst the confusion, Morpheus manages to escape. And what does he do? He jumps out of the shattered windows. The distance to the helicopter is long; there is a seemingly bottomless depth. And yet . . . he takes a leap of faith. He trusts that Neo will grab him.

Morpheus believes that Neo is the One; he believes that Neo will defeat the Agents. He believes that Neo will win the fight against the machines. On what grounds? In all honesty, that basis is rather small. The prophetess—who foretold the coming and the victory—is, in fact, a computer program. The savior—who should bring salvation—doubts himself. In addition, the strength of Agent Smith and the machines is enormous. Morpheus believes without foundation or proof; he believes against his better judgment. Perhaps Morpheus's faith can be best described with the words of the author of the Epistle to the Hebrews. In the eleventh chapter, we read, "Now faith is being sure of what we hope for and certain of what we do not see."

If there is one thing *The Matrix* teaches us, it is that human life consists of leaps of faith. Life does not depend on technology, control, and rationality, but life is about belief and surrender. That is a rather strange message for a science fiction film. *The Matrix* depicts a highly-developed technological society where technology takes over certain important functions from humans, such as sexuality and motherhood, thinking, and decision-making. It is a society where humans and technology converge to such an extent that the faculties of the human mind are amplified by building in hardware and software in the brain. It is a society that can create a virtual world that cannot be distinguished from the "real" world. You would expect that this film preaches some kind of faith in science and technology; you would expect that this film claims that the failures of technology could be intercepted by better technology. But that is not what the film does. On the contrary, *The Matrix* preaches putting faith in prophecy and in peace.

In an interview, Larry Wachowski says that he and his brother are interested in mythology.[1] They want people to think about the important questions in life, especially about life's biggest question: the meaning of being human. Their message is clear: science and technology cannot answer such questions. We can only discover the meaning of being human through the study of belief and what it means to give one's all.

3. "Kierkegaard reminds us..."

What is the view of the Wachowski brothers on faith? What sources do they draw on? Several clues in the trilogy suggest that they were inspired by the ideas of the Danish philosopher Søren Kierkegaard. Direct references and philosophical parallels between the films and the philosopher's ideas bear this out.

Let us first take a look at the overt references to Kierkegaard.[2] The first captain of Zion who volunteered to help Morpheus in *Reloaded* was Captain Søren of the hovercraft Vigilant. A significant detail is that this young captain dies in a fight with the *sentinels* (robots designed to destroy human life). Kierkegaard also died young.

Another reference is the name of one of the most important leaders in Zion: counselor Hamann. On the title page of his book *Fear and Trembling*, Kierkegaard quotes historian J.G. Hamann. In that particular quote, Hamann states that faith is subjective and personal.

The most important reference can be found in the video game *Enter the Matrix*. In the game, the character *Ghost*, who is perceived as the philosophical mouthpiece of the Wachowski's, refers to Kierkegaard as his most important source of inspiration. When *Ghost* is asked why a single man (Neo) can end the war against an entire race of machines, he replies: "Kierkegaard reminds us that belief has nothing to do with how or why.

---

1   See Chris Saey and Greg Garrett, *The Gospel Reloaded* (Colorado Springs: Pinon Press, 2003) 11.
2   Matt Lawrence, *Like a Splinter in Your Mind* (Malden: Blackwell, 2004) 138ff.

Belief is beyond reason (...) Faith, by its very nature, transcends logic."³ This quote is crucial for understanding the relationship between faith and reason in *The Matrix*. Influenced by Kierkegaard, the Wachowski brothers are suggesting that belief has nothing to do with rational considerations (the "how" or "why" of things), but that faith goes beyond reason.

4. Fear and trembling

In his book *Fear and Trembling*, Kierkegaard examines the nature and character of radical faith. He is greatly worried about the religious climate of his time and characterizes the attitude towards religion as a "clearance sale" where everything can be bought at "absurdly low prices." He holds the opinion that faith is dealt with far too comfortably. He wants to raise the stakes. He wants to go back to the situation where faith is "a task for life."⁴ In Kierkegaard's view, faith starts where thinking ends. That is why believing is something incomprehensible. What you can understand of it is "fear, need, and paradox."

What is faith? In *Fear and Trembling*, Kierkegaard offers an extensive analysis of the story of Abraham who is commanded by God to sacrifice his son Isaac on Mount Moriah (Genesis 22). He raises a great number of questions: What did Abraham tell his wife Sarah? How did he feel during the journey? What thoughts and emotions did he have? And, above all, how did he experience this test of faith?

Kierkegaard emphasizes that Abraham believed God's promise that his offspring would be a great people; he held on to that promise before and during the journey to Moriah. Abraham did not believe for the "future," but he believed for "this life." He believed that the promise of God would become reality in his life. And then came this command of God. Did not the command to sacrifice his only son—whom he and his wife had received when they were both infertile—mean he had to give up his greatest wish? Yet, Kierkegaard points out, in giving up this wish, Abraham still believed that God would keep his promise. He thus writes about Abraham: "For it is great to give up one's wish, but it is greater to hold it fast after having given it up, it is great to grasp the eternal, but it is greater to hold fast to the temporal after having given it up."⁵

Kierkegaard distinguishes two movements in the radical Christian faith. The first movement is that of resignation and surrender. It is the

---

3   Quoted from Lawrence 2004, 139.
4   Søren Kierkegaard, *Fear and Trembling*. Translated by Walter Lowrie. Published by Princeton University Press, 1941. sorenkierkegaard.org/texts/text6a.htm.
5   sorenkierkegaard.org/texts/text6a.htm.

movement of "renouncing" and "giving up what you hope to keep." We clearly see this movement in Abraham. He was prepared to give Isaac up; for God, he was prepared to give up his only child through whom he would be able to have many offspring. But in Abraham we see more than just resignation in and surrender to the will of God. And that brings us to the second movement, that of faith. It is astonishing that Abraham, when he appears to be losing Isaac, still believes that Isaac will not be lost. It is characteristic that at a time when all hope of a great people is lost, he continues to believe that his lineage will be great. Abraham expects the unexpected.

Kierkegaard describes these movements: "He mounted the donkey, he rode slowly along the way. All that time he believed—he believed that God would not require Isaac of him, whereas he was willing nevertheless to sacrifice him if it was required. He believed by virtue of the absurd; for there could be no question of human calculation, and it was indeed absurd that God who required it of him should the next instant recall the requirement."[6]

5. Knights of faith

Let us return to Morpheus once more. Kierkegaard's analysis clearly shows what the role of faith in *The Matrix* is. Again and again, we see Morpheus making both movements of faith: firstly acceptance of the situation, and next the faith that "transcends" that situation. We will give two examples.

At the start of *Reloaded* Morpheus is told that the machines are on their way to breaking through to the city of Zion. They are almost within a mile of the gates and they are digging at a speed of almost a mile per hour. Everything suggests that Zion is lost. Morpheus accepts the situation, yet he still professes his belief that "the prophecy will come true."

At the end of the film, the Nebuchadnezzar, Morpheus's ship, is destroyed. The destruction shows the gravity of the situation: the machines are going to win. He resigns himself to this loss by saying that he dreamed a dream that has now gone from him. Despite this, he keeps believing in the fulfillment of the prophecy.

Morpheus is—to use a term from Kierkegaard—a knight of faith. Like Abraham, Morpheus had to make choices in an unclear situation. In this context, Kierkegaard points out two characteristics of faith: (a) the subjectivity of faith and (b) the absurdity of faith.

In Kierkegaard's view, faith is subjective. That means that the be-

---

6   sorenkierkegaard.org/texts/text6a.htm.

liever personally has to interpret God's message. We see this in Abraham, who interpreted God's demand to sacrifice his son as God's will. He could have interpreted it differently, namely as the voice of the devil or of an evil spirit imitating God. We see a similar thing in Morpheus's relationship to the Oracle. He interprets her prophecies as a reliable prediction of the future. He could have interpreted it differently, namely as an attempt of the matrix to deceive the freedom fighters.

The second characteristic is the absurdity of faith. Abraham's situation can be formulated as two opposing propositions. The first proposition is that Isaac is lost (resignation, surrender). The second proposition is that Isaac will live (faith). These two propositions are logically contradictory. But we see here that Abraham's faith transcends logic, and we see the same in Morpheus. On a number of occasions he had to conclude that the prophecy would fail, all the while simultaneously believing that the prophecy would come to pass. Only "on the strength of the absurd" can that contradiction be transcended: in faith.

Kierkegaard calls faith a paradox. A paradox beyond thinking, as faith begins precisely where thinking leaves off. In the words of Kierkegaard: ". . . in order to see what a tremendous paradox faith is, a paradox that is capable of transforming a murder into a holy act well-pleasing to God, a paradox that gives Isaac back to Abraham, which no thought can master, because faith begins precisely there where thinking leaves off."[7]

6. "You have to let it all go: doubt, fear and disbelief"

Does faith begin where thinking ends? Kierkegaard's statement is extremely intriguing as it raises the question of the relationship between thinking and believing. In this section we want to explore this question, drawing on the view of reality as developed in Reformational philosophy. In the next section we will take it a step further and look at this question in relation to religion.

In the third chapter of this book, we provided a tool for articulating the diversity and complexity of reality. If we look at reality from different perspectives, we "see" distinct aspects or dimensions. A few of these aspects are the physical, the biological, the psychic, the analytical, the social, the ethical, and the pistic. Fifteen different irreducible aspects are distinguished in total, and each these aspects have a nature and character of its own.

Human beings function in all these aspects as a subject (see chapter 11). This means that every person is directly subjected to the laws and

---

7   sorenkierkegaard.org/texts/text6a.htm.

norms valid for the different aspects. One of the dimensions is the analytical; meaning that we are gifted with the ability to reflect on matters logically. Another dimension is the pistic, meaning that we are all created with the ability to believe. This belief can be directed at the God of Abraham, Isaac, and Jacob, but can also be directed at an idol.

Thinking and believing are two different functions of human beings. This becomes powerfully manifest when we further explore these aspects. "Thinking" is related to making distinctions between concepts, drawing conclusions from premises, and formulating logical arguments. By "thinking" we often mean *logical* thinking. "Thinking" is bound to certain rules or norms. For example, we can say that someone failed to distinguish between two different concepts or drew an incorrect conclusion from stated premises or is reasoning inconsistently. "Believing" is related to trust in God, the certainty of salvation in Christ, and the renewal of life through the Holy Spirit. Important specifics of such faith are verbs like listening, saying "amen," praying, and praising. "Believing," too, is bound to certain rules or norms, as the Bible makes clear.

Perhaps we can clarify this by saying that human beings have different types of intuitions. Intuition here refers to an insight without analysis. "Reformed epistemologists" like Alvin Plantinga and Nicholas Wolterstorff speak of a "basic belief," an immediate conviction or insight that is not deducted from another insight. However, there are different kinds of intuitions. We all understand that people have moral intuitions, immediate insight into what is right and wrong about a certain action. We can also have the intuition that a certain work of art is well-executed: it is beautiful or ugly or poignant with meaning. But one intuition cannot be derived from or reduced to the other. People have logical intuitions, for example as to whether an argument is correct, a judgment is clear, or presumptions true, which as intuitions differ in kind from the intuitions of faith, like sensing the presence of God, experiencing a glimpse of life's meaning and purpose, or the reality of surrender, devotion, and trust. It has been said that every scientist has basic beliefs (e.g., in the truth of his assumptions), but that doesn't mean that belief and theoretical thinking can be derived from each other. They are, one could say, different cognitive functions. Therefore, Reformational philosophers like Henk Geertsema and André Troost have gone to some length to clearly distinguish the experience of believing on the one hand and theologizing about faith on the other. Believing is *sui generis*.

Not only *The Matrix*, but also the six *Star Wars* science fiction films show the unique character of faith. The characters believe in "The Force," not the biblical God, as a fundamental spiritual entity: "The force is what gives the Jedi his power. It surrounds us and penetrates us. It binds the galaxy together." Faithful followers of the art of living in balance with "The Force" are Jedi masters Qui-Gon Jinn, Yoda, and Obi-Wan Kenobi. But they often meet with skepticism, if not from the average citizen, then from a young student or a fellow warrior. When Obi-Wan Kenobi tries to pass the sentiment for "The Force" to Luke Skywalker, their pilot Han Solo taunts, "Kid, I've flown from one side of the galaxy to the other, but I've never seen anything to make me believe there's one all-powerful force controlling everything. There's no mystical energy field that controls my destiny." Meanwhile, the Jedi are no fools either: they are experts in all kinds of science and technology and are guarding peace and justice in the *Star Wars* universe through their insight into the specific logic of "The Force."

This analysis shows that "thinking" and "believing" each have their distinctive nature and character. More popularly expressed: thinking is not the same as believing, and believing is not the same as thinking. This becomes all the more clear when you try to apply the norms for thinking to believing and the other way around. In the previous section we summarized Abraham's situation in two propositions: (a) Isaac is lost, and (b) Isaac will live. Through thinking, this paradox cannot be "comprehended." That is to say, these propositions are at odds with the laws of logical thinking, which excludes contradictions. But in faith, you can prevail over this paradox. This prevailing is founded in the knowledge that God will keep his promises and that he makes the impossible possible. And, when looking at it this way, one can indeed say with Kierkegaard that believing (in trust) begins where thinking (trying to prove) ends. This does not mean that in believing, the thinking *ends*. Abraham and Kierkegaard, Neo and Plantinga, Obi-Wan Kenobi and Geertsema are no muddleheads. Those who believe, also use their minds.

In this connection, we would like to make one last important observation. Since the Enlightenment, it has become common practice to say that to be human is to be a rational creature, meaning that logical thinking is *the* distinctive quality of human beings. In our opinion, this characterization is flawed. There is no reason whatsoever to give primacy to thinking when characterizing humankind; we all think, believe, create, and evaluate. The distinctive quality about being human is this: we are created in God's image and function integrally as a subject in many different dimensions of earthly existence.

7. "He's starting to believe!": Thinking, believing, and religion

Human beings function in different dimensions. We can make this very tangible: I think, I do business, I enjoy art, I make moral judgments, and with all my heart I believe this or that. When putting it like this, one obvious question is: Who am I? Who is that entity? There is no way you can comprehend this question through thinking. It is—to use Mark Rowland's terminology—a "strange" question (see chapter 11). The question "Who am I?" is, in the end, a religious question. It is a question that we should answer given what we know about God's point of view.

Christians believe that they are made in God's image. They believe that they must aim all their activities at the Creator. They believe that all their actions must be in accord with the laws and norms that are implanted in creation. That kind of believing faith, therefore, affects thought, business, art, morality, and faith-life.

In Reformational philosophy, the image used to describe this is that of a tree with roots and branches. Religion is the root that draws all nutrients from the soil and transports them to the branches of thought, business, art, morality, and faith. Religion feeds all human functioning.

Thereby, religion has, to put in differently, a normative effect on human thought. This conviction has been addressed by philosopher Nicholas Wolterstorff in his book, with the telling title, *Reason within the Bounds of Religion*. He points out that every scientist, when formulating theories, remains within a certain framework of convictions that defines which convictions are acceptable and which are not. And that within that framework of demarcating convictions are also always ontological, metaphysical, and philosophical presumptions.[8]

We ended the previous chapter with Agent Smith's question to Neo during the final great fight: "Why keep fighting?" Why should we as a society keep fighting for a technology that seeks to serve rather than subject? Why keep fighting for a just society? We see no other answer than Neo's: because we choose to. And that choice is—as the analysis above demonstrates—in the end religiously motivated.

---

8   *Reason within the Bounds of Religion* (Grand Rapids: Eerdmans, 1984).

CONCLUSION

Trinity: "It's the question that drives us."
(*The Matrix*)

The subject of this book, according to the preface, is the intersection of three themes: the technological culture in which we live; the science fiction films that give expression to that technological culture; and insights from Reformational philosophy that can help us analyze, evaluate, and regulate technology. We reflected on science fiction and the way in which technology finds its place in that context, and on philosophical questions that can be raised in relation to aspects of a technological culture. It is not a book that will provide much news to a philosophical specialist. Rather, it is a book where Christians join in debates on important issues as they seek to determine and proclaim their attitude. If the reader has kept with us thus far and discovered many interesting, challenging, and informative things along the way, then we have achieved our goal.

Looking back on the previous chapters, there are a number of issues that seem to recur. First, we think it has become clear that a good film—whether it belongs to the science fiction genre or not—deserves time and attention from Christians. They reflect an image of God's reality that more or less depicts what humankind is made of. If only because loads of young (and old) people these days are thoroughly influenced by image culture and popular media, it is important to try and separate chaff from wheat and to approach the practice of producing films as well as consuming films with a well-considered Yes or a critical No—and doing so in a detailed way.

Science fiction films in particular present an image of reality where technology is magnified to such an extent that the assumptions, beliefs, and consequences of a technological culture come to the surface. In our time it is necessary to see which way the wind blows in the development, implementation, and usage of such new technologies. Medical possibilities, for example, are often brought forward in such films to make us think about the desirability of the ever increasing grip of medical research as we think about health and happiness. Technologies shape our images of reality, form our impressions of illusion and reality, and determine our understanding of freedom and control.

Worldview and mindset interweave with our daily work and so often infuse life's means and materials with the desire to reshape the human horizon with new gadgets and megaprojects. Is technology philosophically and morally neutral? We don't think so. There are utopian and dystopian "teachings" blowing in the winds borne by thoughts about technological developments. That is why it is extremely relevant to analyze, evaluate, and regulate technologies from the Christian philosophical perspective.

In this book we drew on a few notions from Reformational philosophy. We spoke about the religious determinedness of technological practices and the direction that a structural given of creation can take (including artistic expressions like film making). We spoke about the many-sidedness of creation. In doing so, we saw something of the specific nature of intelligent machines (intelligent yes, but machines nonetheless) and of the moral responsibility of humankind. Finally, we have looked at the specific nature of that uniquely human reality of faith. Here, too, we have, in relation to *The Matrix*, explored the distorted image of a trusting faith that technological rationality presents, and we sketched in an alternative view from a Reformational perspective.

This discussion is far from complete and far from deep enough, that is true. We began our book with the question: where can we find out what it means to be truly human and how can we know what fits well with our flourishing and what alienates us from the same? Which techniques and technologies are compatible with what is good for a human being? Naturally, an all-sufficient answer was not forthcoming, but we did discover some things as to why an answer is very hard to arrive at. Still, we believe we have come a long way toward deeper reflection on the matrix code. And we invite the reader to continue reflecting on these matters—which can be both constructive and entertaining—with us.

BIBLIOGRAPHY

Film

Roy Anker, *Catching Light: Looking for God in the movies*. Grand Rapids: Eerdmans, 2004.

Judith Barad, Ed Robertson, *The Ethics of Star Trek*. New York: HarperCollins, 2000.

Laurent Bouzereau, *Star Wars: The annotated screenplays*. New York: Del Rey, 1997.

Christopher Grau (ed.), *Philosophers Explore The Matrix*. Oxford: Oxford University Press, 2005.

Richard Hanley, *Is Data Human? The metaphysics of Star Trek*. New York: Basic Books, 1997.

William Irwin (ed.), *The Matrix and Philosophy*. Chicago: Open Court, 2002.

———, *More Matrix and Philosophy*. Chicago: Open Court, 2005.

Ross S. Kraemer, William Cassidy, and Susan L. Schwartz, *Religions of Star Trek*. Oxford: Westview, 2003.

Stephen Law, *The Philosophy Gym*. London: Review, 2003.

Matt Lawrence, *Like a Splinter in your Mind: The philosophy behind The Matrix*. Oxford: Blackwell, 2004.

Mark Rowlands, *The Philosopher at the End of the Universe: Philosophy explained through science fiction films*. London: Ebury, 2003 (previously published as *Sci-Phi. Philosophy from Socrates to Schwarzenegger*).

William D. Romanowski, *Eyes Wide Open: Looking for God in popular culture*. Grand Rapids: Brazos Press, 2001.

Chris Saey, Greg Garrett, *The Gospel Reloaded. Exploring spirituality and faith in The Matrix*. Colorado Springs: Pinon Press, 2003.

Glenn Yeffeth (ed.), *Taking the Red Pill: Science, philosophy and religion in The Matrix*. Dallas: Benbella Books, 2003.

Reformational philosophy

Roy Clouser, *The Myth of Religious Neutrality.* 2nd ed. South Bend: University of Notre Dame Press, 2005.

Stephen V. Monsma (eds), *Responsible Technology*, Grand Rapids: Eerdmans, 1986.

Richard J. Mouw and Sander Griffioen, *Pluralisms and Horizons.* Grand Rapids: Eerdmans, 1993.

Alvin Plantinga, *Warrant and Proper Function.* Oxford: Oxford University Press, 1993.

Egbert Schuurman, *Faith and Hope in Technology.* Toronto: Clements Publishing, 2003.

Sytse Strijbos, Andrew Basden (eds.), *In Search of an Integrative Vision for Technology: Interdisciplinary studies in information systems.* New York: Springer, 2006.

Maarten J. Verkerk, *Trust and Power on the Shop Floor.* Delft: Eburon, 2004.

Albert M. Wolters, *Creation Regained: Biblical basics for a reformational worldview.* 2nd ed. Grand Rapids: Eerdmans, 2005.

Nicholas P. Wolterstorff, *Keeping Faith: Talks for new faculty at Calvin College.* Occasional papers from Calvin College, Vol. 7, No. 1, Grand Rapids: Calvin College, 1989.

———, *Reason within the Bounds of Religion.* Grand Rapids: Eerdmans, 1984.

———, *Divine Discourse: Philosophical reflections on the claim that God speaks.* Cambridge: Cambridge University Press, 1995.

www.ingramcontent.com/pod-product-compliance
Lightning Source LLC
LaVergne TN
LVHW091300080426
835510LV00007B/336